D1137559

The Mystery

Matthieu Ricard

# of Animal Migration

English version by Peter J. Whitehead

Paladin

Granada Publishing Limited
Published in 1971 by Paladin
3 Upper James Street, London W1R 4BP

First published in Great Britain by Constable & Co. Ltd 1969
Copyright © Robert Laffont 1968
This edition Copyright © Constable & Co. 1969
Made and printed in Great Britain by
Hazell Watson & Viney Ltd, Aylesbury, Bucks
Set in Monotype Ehrhardt

This book is sold subject to the condition that it shall not,
by way of trade *or otherwise*, be lent, re-sold, hired out, or
otherwise *circulated* without the publisher's prior consent in
any form of binding or cover other than that in which it is
published *and without a similar condition including this
condition being imposed on the subsequent purchaser*.
This book is published at a net price and is supplied subject
to the Publishers Association Standard Conditions of Sale
registered under the Restrictive Trade Practices Act, 1956.

Contents

List of Diagrams

Introduction

What do we mean by migration? The word usually conjures up the idea of a long journey, possibly because most of us associate migration with such long-distance migrants as swallows, which travel thousands of miles during their migrations. One should, however, bring the notion of *periodicity* into a definition of migration and not conclude that a migration must necessarily be defined in terms of the actual distance travelled.

Migration, then, is essentially a regular movement in space, no matter what the extent of the displacement may be. In discussing examples of animal migrations, however, one must also distinguish between true migration and *invasion* (or mass exodus), the latter being a migration without return. Migration really involves a loop, so to speak, both in time and in space, and one can thus distinguish a definite outward and return journey, the two defined with regard to the place in which breeding takes place. The migratory cycle most often takes place within the span of a single year, which is what one would expect since the year, like day and night, is a very basic unit in the rhythm of natural events.

This does not, of course, exclude more complex cycles in which the period is more or less than a year; our definition merely states that a migration must embrace some regular cycle.

Having defined migration, we can now proceed to the various examples of it in the animal kingdom, and demonstrate that, in spite of the wide diversity of animal forms, there are some basic features of migration, navigation and orientation that are common to a host of different groups. The world of birds is particularly rich in migrant species and it has seemed instructive to treat this chapter more fully than the others; also, birds have been rather better studied than most other animals. A fairly large number of species has been mentioned because birds are, on the whole, easily observed and more species are probably known to the non-specialist than is the case, for example, with fishes.

Mammals

The migration of certain animals offers some of the finest spectacles that nature can provide. Regretfully, however, one must often say 'has offered' because migratory mammals, even more than migratory birds, have been decimated by man, and some species have been reduced to total extinction. The larger mammals, which are often excessively vulnerable when in herds, have been shamelessly exploited by men blinded by the power of new means of destruction. In this respect, the period that we shall deal with is one that is marked by a succession of massacres: massacres of seals in the rookeries of Alaska, massacres of whales, of antelopes and caribou and the slaughter of nearly 100,000,000 bison.

The American Bison (*Bison bison*) is symbolic of many things. It is a symbol of strength: its massive bearing, thick-set neck, heavy head and obstinate brow, together with its weight (often exceeding two tons), all give the impression of strength and power. Again, its herds are a symbol of the colossal forces of nature, like a cresting wave, a cyclone or an equinoctial tide. But the bison is yet another symbol – a symbol of the senseless destructiveness of modern man. Well-informed people are not altogether surprised to see here again one more example of man's thoughtlessness that is so apparent in other domains. True, the bison has survived the total extinction meted out to some other victims of this modern world. But this is surely only by chance, for of the 70,000,000 believed to have existed before the Europeans appeared on the prairies of North America there were only 1,900 bison left in 1889. Normally, when the numbers of breeding animals fall below a certain level, the chance of individuals mating outside their immediate circle of close relatives is reduced, natural selection fails to operate properly, the number of sick and genetically unfit animals increases, and epidemics can soon finish off the rest of the population. However, thanks to energetic action by the American government, the bison was saved in the nick of time.

We shall draw mainly on the work of Martin S. Garretson, a specialist in this field, in the following discussion. He estimated that there were 21,500 head of bison in 1933. Although it is difficult to give as accurate a picture of bison migrations as would be possible nowadays with proper methods of observation, nevertheless the size and extent of the phenomenon were such that even from travellers' accounts and isolated observations a fairly clear idea can be gained. The bison occupied a large part of North America, from Mexico to Canada, mainly in the plains and prairies lying to the east of the Rocky Mountains, and the population must have numbered about 70,000,000. A certain part of this population, perhaps 5,000,000, also colonized the mountain valleys, for the bison, contrary to expectation and to popular opinion, is extremely agile in mountain country. Outside the migratory periods, the huge herds would break up into small tribes which would wander in search of pasture. The migrations seem to have been activated by the search for food, although, like some other migratory animals, the bison nevertheless sometimes left rich pastures or a warm environment while at other times they passed the winter in the snow, surviving biting cold winds and a dearth of food.

The main axis of these migrations was north–south. In spring those bison that had wintered in the north of Mexico and the plains of the south-east United States pushed northwards as the new shoots of grass began to cover the prairies. After the summer rest and with the dry days of autumn and the cold of winter approaching, the herds would begin to move southwards.

An early account of bison herds was given by Townsend in his *Narrative of a journey across the Rocky Mountains* in 1833–4:

Towards evening on the rise of a hill, we were suddenly greeted by a sight which seemed to astonish even the oldest among us. The whole plain, as far as the eye could discern, was covered by one enormous mass of buffalo. Our vision, at the very least computation, would certainly extend ten miles, and in the whole of this great space, including about eight miles in width from the bluffs to the river bank, there was apparently no vista in the incalculable multitude. It was truly a sight that would have excited even the dullest mind to enthusiasm.

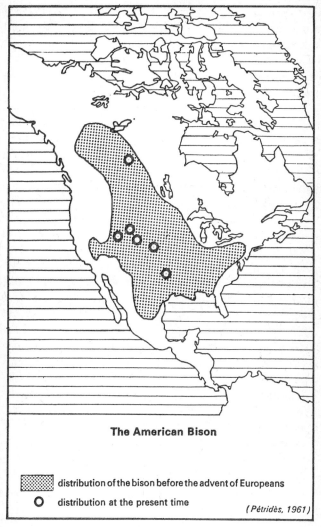

The American Bison

distribution of the bison before the advent of Europeans

O   distribution at the present time

(Pétridès, 1961)

Fig. 1

In Arkansas in 1871, between Fort Zarah and Fort Larned, a compact herd of 4,000,000 bison was seen and it seems that such a sight was in no way exceptional. In addition to these, there are many other estimates of the sizes of the herds, some of them almost unbelievable. General Sheridan and some of his staff once tried to estimate the numbers in a herd between Fort Dodge and Fort Supply. Their first estimate was ten milliards (10,000,000,000), then 1,000,000,000 and finally they settled for a 100,000,000, but this they felt to be an under-estimate.

Few things could stop such a mass of animals on the move and the bison were simply not able to retrace their steps if the way ahead was blocked, so great was the press of animals behind.

Once, during a snow-storm, a herd of not less than 100,000 bison crashed over a steep cliff. The leaders, in order to face up to the storm, had left the usual trail and came to the edge of the precipice. Forced on by those behind, they tumbled over. Those following were in the same predicament so that eventually there were some thirty feet of solid bones at the foot of the cliff to bear witness to this migration.

The migrations were variable in extent and it is doubtful that the same herds that spent the summer in Canada were those that wintered in Mexico. It is thought that there was an over-all shift of the bison herds some 300 miles southwards in winter and then a drift back again of 300 miles northwards during the breeding season. East–west movements, while limited by the Rocky Mountains, certainly took place and a part of the herds moved into the foothills.

Year after year the trails were the same. The tracks formed by successive generations were remarkable in many ways, notably for following so closely the contours of the land, to such an extent that when the survey parties came to align the new railway routes, they often used the old bison trails because they involved the gentlest gradients. Soon, however, the trails were disrupted, the herds decimated and the majestic migrations of the bison passed into legend. Impassioned pleas against their slaughter were of no avail. From the time of the Indians the herds had lived in perfect equilibrium with their surroundings, with few natural enemies, only the sick or aged falling prey to the coyotes (thus ensuring the general well-being of the herd). The arrival of the Europeans changed all that.

The slaughter began. First, the bison were killed for their meat then only for the tongue which was considered a particular delicacy (the remaining two tons being left to rot). Then they were killed in order to deprive the Indians (who in their turn were slaughtered) of their means of subsistence. Finally, they were killed for 'sport'. There were competitions, even a world championship (from an old poster):

Grand Excursion to Fort Sheridan
Kansas Pacific Railroad
BUFFALO SHOOTING MATCH
for $500 a side and the Championship of the World
between
Billy Comstock (the famous Scout)
and
W. F. Cody (Buffalo Bill)
Famous buffalo killer for the Pacific Railroad

Buffalo Bill Cody, who had already killed 4,820 bison in eighteen months, roused the enthusiasm of the crowd by shooting sixty-nine bison in the course of this contest. Comstock shot forty-six.

Between 1870 and 1875 not less than 2,500,000 bison were slaughtered annually. In some parts of Kansas there were places where the plains were whitened by bones as far as the eye could see.

As a contrast to the bison we shall now turn to the Springbok (*Antidorcas marsupialis*), one of the most graceful of creatures, as much in its appearance as in its behaviour. It gets its name from its habit of leaping vertically, when frightened, to a height of about eight feet and landing on almost the same spot and at the same time raising the white hairs along the hind part of its back into a kind of 'fan'. A herd of frightened springboks is a wonderful spectacle as they perform their graceful 'ballet'. Indigenous to South Africa, these gazelles were once extraordinarily abundant. But with the growth of the modern world the same fate befell them as that of other animals that form vast herds. Although they have not entirely disappeared, their numbers are greatly reduced.

In the old days scientific and objective study was not considered necessary, so that stories of the springboks and their migrations, although poetic and beautiful, do not enable us to give an exact account of the phenomenon. Nowadays, of course, such a study is impossible, because the herds have been so drastically reduced and the former conditions have altered. Nevertheless, the accounts that we do have give us a fairly clear general picture. In the first place, it seems that two types of migration were involved.

The first was an annual migration correlated with the natural rhythms of the seasons (i.e. the variations in temperature and rainfall which affect the growth of plants and grass on which the springboks feed). The second type of movement was a sort of 'lemming phenomenon' (see page 24) in which millions of springboks began to march, inexorably, throwing themselves into rivers or the sea, having suffered attacks by predators running in their midst (wild dogs, jackals, hyaenas and other wild animals).

The first type of migration was an annual movement closely tied to the climatic and reproductive cycles. In winter the springboks went up on to the high ground of south-western Africa, retreating before the spread of dry conditions from the east. At this time there is more rain in the hills than elsewhere and here the young calves were born and nourished on new and tender shoots. Later, the springboks would return to the north and east where the rains had by then restored the grass and filled the pools. But if these rains failed, then the animals sensed it (or rather detected it in some mysterious way) and remained in the highland areas. The distances travelled during these migrations were often considerable.

The second type of migration was essentially a mass exodus, the 'treks' as the Boers called them and legend records. The treks occurred irregularly, usually every two, three or four years, and the springboks then completely altered their normal behaviour. Formerly shy, lively, leaping, they would now advance in compact ranks pacing slowly and relentlessly forward, fearing neither man nor beast, barely noticing men who might walk through the herd to get an even better view of this extraordinary phenomenon. Nothing but springboks was to be seen for miles and miles, so that the entire countryside appeared to be in motion. From 1887 to 1896 there were four large treks of this kind, three to the north and one to the south-west. The following is an eye-witness account of such a trek in July of 1896:

While crossing a small rise in the ground we were faced with a vast plain, peaceful and resplendent. In a single sweep our eyes embraced a huge stretch of brown earth dotted with small hillocks into the distance and bathed in the marvellous and brilliant tints of the Karoo; and across this vista the delicate antelopes quietly browsed under the midday sun. We three farmers were used to estimating numbers in small herds and we possessed an excellent pair of field-glasses. I suggested to my friends that we might try to make a correct estimate of the number of springboks in front of us. With the aid of our binoculars we made a very careful estimate taking them by sections and making a cross-check on our calculations. In the final count, we reckoned the number was at least 500,000, a half a million springboks in view at one time. I do not hesitate to say that this estimate was not excessive. We were quite accustomed to the vast plains of Africa and the views that they afforded, but now we remained silent, feasting our eyes of this wonderful spectacle. Now to give an idea of the prodigious number of antelopes comprising this herd one must take into account that it was estimated to stretch for a journey of twenty-three hours in one direction and two to three hours in the other, which meant that the herd occupied an area of 140 by 12 miles. Naturally the herd was not equally dense throughout, but when one says that there were millions present this is literally true.

The causes behind this extraordinary behaviour on the part of the springboks which drives them in their millions to destroy a large part of their population are the same as those that drive the lemmings into the sea, and those behind the migrations of locusts. For there is no doubt that this collective sacrifice, which solves the problem of competition for food and restores the natural balance, is deep-rooted in the instinctive behaviour of the species.

Some believe that the motive behind these migrations is the search for water, for one certainly sees the springboks throw themselves in their millions into the waves and drinking the sea water that quickly kills them. The beaches are then crammed with innumerable corpses which are washed up again at each tide. But if the springboks often start these treks looking thin and emaciated, there are other occasions when they begin fat and sleek, which rather weakens the argument that it is the search for water that is the prime cause. The treks are migrations without a return, the herd being worn down rapidly by the obstacles encountered, the lack of food and by predators until, like melting patches of snow, the herd finally vanishes. Today these great treks are a thing of the past. The population of springboks has become so depleted that the mechanism triggering off these suicidal migrations no longer has reason to operate.

There are certain other African antelopes that are migratory, such as the Great Kudu (*Strepsiceros strepsiceros*), but there is nothing nowadays to compare with the former treks of the springboks.

Europe, however, possesses an antelope that travels great distances during its annual wanderings. This is the Saiga Antelope (*Saiga tatarica*), a beast with a massive body, thick-set legs and a bulbous and curiously prominent snout which gives it a rather droll appearance in comparison with the delicate elegance of the springbok.

In prehistoric times the saiga was widely distributed across Europe, reaching even as far as the foot of the Pyrenees. Having almost disappeared as a result of intensive hunting, with only a few hundred individuals left at the beginning of the century, the saiga has now prospered and the population comprises not less than 2,000,000 head. The saiga now occupies some 1,000,000 square miles of steppeland to the east of the Don. They spend the winter to the south of the 46th parallel, retreating before the advancing snow but in summer returning to their northern pastures and thus covering huge distances.

Another animal that was once very much more widespread than it is today is the African elephant. But the value placed on its ivory made it an unwilling contender in a merciless war. After a marked decline strict protection measures have succeeded in raising the number of elephants in some areas – occasionally too high it would seem, if one is to preserve a natural balance between all the different species in a park or reserve. The area now occupied by the African elephant, however, is still very much as it was before protection. Elephants also migrate and their movements over the years are so regular that the trails pounded by them remain bare of vegetation. In the hot season the elephants retreat into the forests, but when the rains break the high humidity forces them out into the scrub areas. Elephant trails, always well known to the local inhabitants, are often hundreds of miles long.

The numerous members of the deer family (*Cervidae*) that are found in most parts of the world are variously migratory. With the exception of the reindeer and the caribou, which are habitual migrators, the remainder tend to move only when climatic conditions become unfavourable.

The Wapiti or North American Deer (*Cervus canadensis*) has, thanks to the large reserves made for it by the American government, a sufficient degree of freedom, and a sufficient range, for useful studies of its movements to be possible. It is particularly in mountainous regions, where temperature and other seasonal variations are most marked, that the most regular movements occur. In summer the mountains offer the deer the best pasturage but not when the slopes are covered by snow. The migrations of the herd in Yellowstone National Park (some 20,000 head) best illustrate the extensive pattern of 'transhumance' (a term usually applied to the seasonal shift of agricultural activities in the mountains of Switzerland but one which seems appropriate here). The wapiti descend to the valleys as the first snows arrive, shortly after the nuptial assembly. The spring migration is in the reverse direction. Year after year the wapiti return to the same locality that they left, rarely abandoning the places they already know. To reach their particular area they follow the ancestral tracks, which have not varied within the memory of man. Only during especially hard winters when there have been greater falls of snow than usual will the deer search out better places. The descent to the valleys is often complicated by the complex movements of the nuptial assembly, when the groups break up and exchange members. The assembly over, the groups make for their respective winter quarters, and the result is a curious criss-cross pattern of migration, and sometimes one sees the columns of deer going off in several different directions.

In eastern Europe the tentacles of civilization have created barriers across the natural environment, just as the completion of the Union Pacific Railroad in 1865 cut the bison herds into two groups. Herds of deer that formerly roamed are now confined to narrow 'islands'.

The building of highways flanked by fences and the lining of the canal banks with concrete (animals trying to cross often drown, being unable to scramble up the smooth, steep banks) now prevent the movements of the larger mammals such as deer and wild boar, and in this way the natural equilibrium is seriously upset. In the Swiss National Park, however, one still finds regular migrations of a dozen miles or so between summer and winter pastures.

The Reindeer (*Rangifer tarandus tarandus*) is certainly one of the last of the larger mammals whose migrations can still be seen. In Europe there is a wild population on the islands of Scandinavia and in Siberia, and a semi-domestic population in Lapland which is part-guided, part-accompanied by the Lapps. On Spitzbergen the reindeer spend the summer months on the grassy plains from which the ice and snow have melted, and then during the autumn they descend to the coasts where they live on seaweed washed up by the tides. When winter comes they return to the higher ground in spite of the intense cold and they then scrape away the snow and browse on lichen and other plants that somehow manage to grow beneath the surface. When they descend once more to the coast in the following spring they are fat and sleek. It is at that time that they face the most hazardous part of their migrations. The sun is still weak and has not completely melted the snows, so that the slopes are slushy at midday but form icy sheets at others times.

The populations on the mainland have more space in which to roam. In eastern Siberia herds of thousands of reindeer descend from the mountains in the north and, crossing the plains, make for the taiga forests where they spend the winter. In the spring, wreathed in great clouds of mosquitoes that have hatched from the now swampy forests, the reindeer make their escape to the higher ground once more. Like all large migrations, the movements of these huge herds have occasioned many splendid descriptions: such a display of the vast natural forces in our environment is indeed a rare but unforgettable sight. In Lapland the lives of the Lapps are closely tied to the habits of the reindeer, providing a simple but excellent example of the way that man can use nature without destroying it. The Lapps carefully follow the migrating herds, playing their part in natural selection by appropriating the weaker members of the herd and sometimes altering the path of the migration. Quite often when the reindeer have reached the shores, the Lapps lasso the leader and pull him into the water. After some hesitation the remainder of the herd follow, and thanks to this strategy the whole troop soon cross over to one of the offshore islands.

The crossing, which might take an hour, is not without

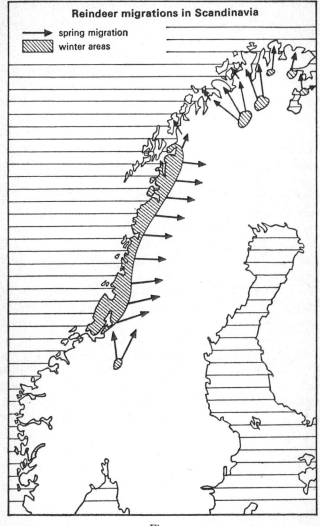

Fig. 2

dangers for the younger animals, and these must then be helped by men who lash them to their boats. In the mountainous islands the reindeer find rich pastures and a refuge from the mosquitoes. In autumn the herds are assembled and, taking advantage of calm conditions, they recross the strip of sea and press on into the interior.

The Caribou (*Rangifer tarandus arcticus*), as the reindeer is called in America, is now the largest of all the migratory mammals. As one might expect, the destructive forces in our modern world have dealt harshly with the caribou and of the 100,000,000 that must have once existed not more than 300,000 now remain. But unlike other large migratory mammals, they still pass through those huge areas of taiga and tundra that even now have not yet been upset or modified by the hand of man. In winter, forced southwards by the extreme rigours of the Canadian climate, the caribou take refuge in the taiga. The southern limit reached in the course of their wanderings is approximately on the north-east corner of Lake Ontario. The taiga is the northern forest, essentially coniferous (deciduous larches and evergreen pines and firs) but also containing maples. The taiga forests are fairly open, for there is no undergrowth, but only a thick carpet of lichens and mosses on which the caribou feed. Confronted by deep layers of snow, the caribou find food by scraping the ground with their large hooves. Unfortunately the taiga is often ravaged by forest fires which play havoc with the natural balance of life. Fire not only destroys the vegetation, but kills off animals, destroys the lichen (which takes a long time to regenerate), and also destroys the bacteria that are responsible for maintaining the fertility of the soil; worse, it accelerates soil erosion and generally makes the ground barren. After the cold winter the caribou begin their northward migration, reaching a peak in April–May. Trotting, sometimes half-galloping, the herds merge together until many thousands are moving northwards in stages of about ten miles a day.

From the taiga they move on to the tundra, which Fred Bruemmer has described thus:

Lying to the west of Hudson's Bay, between the taiga and the Arctic Sea, [the tundra] occupies almost twice the area of France. Dissected by tens of thousands of lakes, of which only the larger have received a name, traversed by large rivers, it is a countryside at once monotonous in general appearance yet varied in detail; here, a bare rocky expanse where only mosses and lichens flourish, there, only a few miles away, a prairie covered with grasses and flowers.

These are the barren grounds, the dead lands, which give their name to the Barren Ground Caribou, as this subspecies is called.

The caribou behave in a particularly intelligent way towards obstacles in their path. When they are confronted with a frozen pool or lake the entire troop halts and they walk across in Indian file, carefully stepping across a slippery surface that might well give way were the whole herd to cross at once. When the ice breaks up, they swim rivers after having found the shortest crossing and then, row upon row, a huge triangle moves across with the leader at its apex. After a vigorous and often amazingly rapid swim they reach the other shore.

And then suddenly it is summer. The silent mornings of spring give way to the chatter of bird song, the streams pour down in torrents beneath the still unmelted ice, myriads of flowers one day suddenly burst forth on the carpet of lichen, and a million mosquitoes set up their exasperating hum. It is June, and the female caribou separate from the herd to bear their young in some safe place.

Soon the young caribou, who can walk and swim after only a few days, join the migration northwards, accompanied by their mothers who protect them from the ravages of wolves and other predators. During the short northern summer the caribou cross and recross the tundra in all directions searching for rich pastures. But all too soon there is a nip in the air, the cold weather sets in, and although protected by their thick coats the caribou more often prefer to retrace their steps. The icy weather reminds the herds that they have tarried too long and the last ones finally leave the taiga towards November. Good year or bad year, the caribou thus make a round trip of many thousand miles along their age-old tracks.

If the barren ground caribou lives under difficult conditions, what can be said of the Woodland Caribou (*Rangifer tarandus caribou*), a subspecies that scorns the cold and the blizzard by making its way northwards in autumn and scoffing at its relatives as it returns southwards in the spring.

The migrations of the lemming have always aroused the

greatest curiosity. There are three species of these small rodents living in Scandinavia. *Lemmus lemmus*, the Tundra lemming, is the lemming of the high plateaux. It is the prettiest of the three but its most important feature is its size, about twelve centimetres. Its coat is short and a soft beige which tones in with its surroundings. There are two black marks on its back, white whiskers and a pair of small round eyes that are jet black and give it an inquisitive look. Like all small mammals that live in fields, and open country, it is both agile and ferocious and can prove quite vicious when caught. In the late afternoon, after a long siesta, the lemmings leave their burrows and go about discreetly in search of food. Their diet is fairly catholic and they are equally content with seeds or insects. When winter comes the lemmings abandon their burrows and construct a cosy nest of leaves and twigs, generally in a bush, a little distance above the ground. Intense cold and deep snow do not affect their activities and thanks to a network of tunnels burrowed under the snow they are not confined to their nests.

The lemming of the forests (*Myopus schisticolor*) is smaller (about ten centimetres long) and is darker in colour. The Arctic Lemming (*Dicrostonyx torquatus*) is the largest, reaching fifteen centimetres in length. It has a long and shining coat of beige fur enriched with reddish tints and there are black markings on its shoulders and front.

First contact with the lemming reveals an animal that is solitary and retiring, preferring to move around at night and on the face of it an unlikely animal to hold a legendary place amongst the great migratory creatures of the world. But suddenly, one year, the females which normally produce 4 or 5 young in a year start to produce 6 or 8 not just once but 4 or 5 times in the season. It is this that upsets the natural balance – 25–40 young instead of the usual 4 or 5. Such an enormous increase in fecundity is altogether amazing. It invariably happens after a period of especially good feeding, which seems to favour an increased hormone metabolism in the females. But it is not long before the pendulum swings the other way. Very quickly the burrows become overcrowded, the hunting grounds become too small and over-population becomes acute. But instead of letting equilibrium re-establish itself by a combination of famine and the host of predators that descend on this concentration of prey,

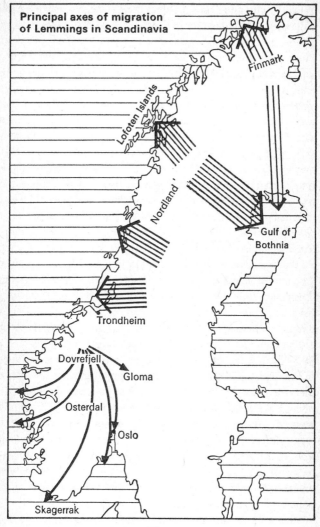

Fig. 3

25

nature takes a different course and provides a much more spectacular solution. This is the mass exodus.

At a certain point the lemmings start to migrate, at first each on his own; but soon the numbers swell, so that although they are trying to escape overcrowding, it is a veritable flood of lemmings that sweeps down from the hills to the plains. Once solitary and timid they now break loose and become quite ferocious in the pursuit of their goal. Nothing stops them. Excellent swimmers, they leap into the water whenever a river or lake bars their way, swimming across such barriers rather than deflect their course for an instant. If a boat happens to be in the way, then they somehow scramble up one side and down the other. If they come to a village, they invade it and passers-by are sometimes bitten on the legs; they enter houses and swarm into the rooms, never making a detour of any kind. But inevitably, one day, having travelled at the rate of some fifteen miles a day, they come face to face with the sea. It does not dismay them any more than the other obstacles; it is their destiny. Relentlessly they throw themselves in their hundreds of thousands into the waves, desperately trying to continue on their self-appointed route. Their incredible determination is such that some of them may reach the Lofoten Islands off the coast of Norway.

Where would they go if they were not stopped by the sea? No one knows. Some believe that before the North Sea came into existence the lemmings must have made regular migrations to the British Isles, and that this primordial instinct still drives them on in this direction (a similar theory concerns the migration of eel larvae – see page 156). But such an explanation must be rejected in the case of the lemmings of Scandinavia because of the many different places from which they start and the different directions in which they go (see map, page 25). The true explanation is much more likely to be found in the results of certain experiments that have been made on the degeneration of social behaviour in animals.

The work of an American, Dr John B. Calhoun, on the behaviour of colonies of rats seems to throw light on the problem. Calhoun studied the effects made by overcrowdings on the social behaviour of rats. Normally, in a colony of rats, a hierarchy is established whereby the strongest member dominates the weaker one and tends to organize the social life of the colony. All goes well, the population becomes adjusted to the space available and the females build normal nests and successfully raise their young. If, however, it is contrived that the population becomes over-crowded, then the social behaviour within the colony rapidly deteriorates. To achieve this, the rats are put into a box or cage with two openings through which they can enter or leave. In this way the leader of the original colony is unable to guard both entrances at once so that rats from another colony can enter. Before long the colony becomes overcrowded and the social hierarchy is upset. As soon as this happens the whole social order degenerates: the rats start to fight for food; they lie in wait for each other and as soon as one makes its way towards a feeding bin it precipitates a rush from them all, other feeding bins being ignored; the females build unsatisfactory nests; the infant mortality rises to over 96 per cent. This kind of social degeneration demonstrates how overcrowding can completely change the whole pattern in the behaviour of a community.

It seems likely that the mass exodus of the lemmings is linked if not to the same mechanism, then at least to something rather similar. Thus collective suicide is a self-regulating mechanism by which harmony is once more restored. There are always a few individuals left to perpetuate the race – until over-population once more brings on this collective sacrifice. Naturally, these huge population explosions spark off a chain of biological events. Predators find a prodigious source of food at this time. The Snowy Owl (*Nyctea scandiaca*) gorges itself with lemmings, and under such conditions raises a full brood of chicks. Normally, this owl lays six to thirteen eggs, but when food is scarce few of the young survive. But if there is an abundance of lemmings, then the chicks get their full share of the food and the population of owls increases greatly. The following year, however, such a population would be too large for the food resources, and this time it is the birds that must migrate. While many must perish, others migrate successfully and make an appearance in places where they are normally not seen, as in the Shetlands in 1967. There is also a very strict relationship between the complement of Canadian Lynx (*Lynx canadensis*) and the Canadian lemmings. There are many species of lemming living in the tundra regions of Canada and these are fairly closely related to those of Scandi-navia described above. During the 'lemming years' (every three, four or five years) the lynx multiply, but the succeeding years bring a harsh reversal and many die off. It is not only owls and lynx that feast to their hearts' content. Skuas (*Stercorarius* species) also have strongly predatory habits, as also sea-gulls, ermines (stoats), weasels, foxes, dogs and cats, and all contribute to reducing the sudden explosion of lemmings.

The main migratory routes of the Scandinavian lemmings are shown on the map on page 25. The interval between these migrations is usually between three and five years, and never more than ten. Since 1941, however, no more large migrations have occurred. This is certainly due to the retreat of wild life before the advance of civilization. Farmers may not be much concerned, for the lemmings do considerable damage to crops, but it will be regretted by those interested in the more spectacular aspects of our rapidly vanishing natural fauna.

It has long been held that bats, like certain birds, are able to make long migrations, just as these creatures, of which the insectivorous European species are quite inoffensive, have also been considered in some way malevolent, leading to many weird and quite unscientific legends. In fact, bats do quite often disappear from a locality in winter-time, reappearing the following year in quite a regular manner. Three types of behaviour are found in bats which lead to this disappearance in winter.

The first and most general explanation is that bats tend to cease all their activities with the approach of cold weather and to go into hibernation until the spring. The second explanation is a small-scale type of migration in the course of which colonies emigrate a short distance from one locality to another in order to find the right conditions of temperature (cold) and humidity under which to hibernate. The distances travelled are of the order of thirty miles. This type of migration is still not properly understood, for the colonies will often leave a cave that is dark, damp and cold, only to take up residence in another where conditions appear to be identical. Again, it is not unusual to find that the place to which they emigrate is either to the north *or* to the south of their summer residence, and in some cases one finds a dispersal in all directions.

The third explanation for the disappearance of bats is a genuine large-scale migration much like that found in birds. Certain American bats illustrate this type. One of them, the Rufous Bat

(*Lasiurus borealis*), which lives and breeds in trees, unlike its cave-dwelling relatives, migrates in autumn from the north of Canada down to the southern-most parts of the United States. Often they pass over the ocean, for these bats have been recorded from Bermuda, and the Golden-eared Bat (*Lasionycteris noctivagans*) has been seen off Cape Delaware, fifteen miles from the shore. The Molasses Bat (*Tadarida brasiliensis*) forms immense colonies in the spring, made up of many millions of individuals which congregate in the maze of passages and caverns in the Texan caves. In autumn the colony divides. The majority of females emigrate to Mexico, whereas the males are sedentary and remain. These caves have been colonized since time immemorial and the floors are carpeted with a thick layer of guano. The guano makes an extremely rich fertilizer and for this purpose wooden towers are built in the United States to harbour the tens of thousands of bats, a single tower yielding about two tons of guano yearly.

Bats can be divided into two main groups, the insectivorous and the fruit-eating species. Many of the latter are found in Africa and their movements can be correlated with the ripening of fruits, whereas in the former type it is often the pattern of insect emergence that determines times of migration. In Australia the fruit-eating Grey-headed Flying Fox (*Pteropus poliocephalus*), which proves such a pest to farmers, migrates from Queensland in the north in October and reaches the southern part of the continent by December. The farmers shoot them in flight and also dynamite the trees which form their 'dormitories', but this seems to make no appreciable difference to the vast numbers of flying foxes.

The European bats are mainly small-scale migrators. Apart from the Mouse-eared Bat which will be discussed below, marking experiments have resulted in two actual cases of long migrations, the first of 900 miles and the second of 1,200 miles, both by bats from Scandinavia.

The mouse-eared bat, as Norbert Casteret has shown, is a most interesting migratory species. Casteret, the well-known speleologist, made a series of highly interesting observations on these bats which have greatly added to our knowledge of their habits. The mouse-eared bat (*Myotis myotis*) is one of the largest of the European bats and one of the strongest flyers. (Contrary to popular opinion, bats are far from being weak and ineffectual flyers, as witness the celebrated observation of bats accompanying swallows.) The young of the mouse-eared bat are born in June in what might be called a collective confinement, the females of the colony all giving birth on the same day. The reproductive cycle in these bats shows odd peculiarities, of which the phenomenon of delayed fertilization is one of the most remarkable. The nuptial assembly and copulation take place in the winter quarters just prior to hibernation. Throughout the winter the metabolic processes are very greatly slowed down and it is only when the bats reawaken from their winter sleep that the sperm which was deposited in the autumn fertilizes the egg. Thus the gestation period begins during the return journey to the summer quarters in April and May, and the embryos reach full term at the end of spring. Such delayed fertilization is also found in other animals in which the metabolism is slowed down during the winter (e.g. the Alpine Marmot, *Marmota marmota*). A similar phenomenon is found in deer and bears in the northern hemisphere, and curiously enough it can be induced in the Malayan Sun Bear when this animal is transferred to a European zoo.

For a month the young mouse-eared bats clutch their mothers, who feed them with milk and insects. Then comes the parting, the mother having become more and more weighed down with her burden. The young one is disengaged and, surprised at this brusque expulsion, it clutches awkwardly at the roof of the cave. The mother continues to bring it food from her nightly sorties and only stops when the juveniles take to the air and join the hunt. The summer over, the time comes for the autumn migration. During the preceding days the bats are seized with a considerable nervous activity. Instead of a comfortable, sleepy community – for bats cling quite easily to the roofs of caves by means of their curved claws, the wings folded to form a hammock for the body – there is now much activity, with bats squeaking and flying in all directions. When night falls, they leave the cave and make for their distant goal.

In general, migrating bats fly at a height of about 150 feet from the ground (sometimes higher) and they prefer to fly at night. If bad conditions have slowed them down, however, they may continue to fly in the daytime in order to reach the caves that they frequent year after year along their route. Such stopping-places seem to be fixed by tradition. The bats of south-eastern

France find their migratory route barred by the line of the Pyrenees, but, guided by the older females, they succeed in crossing them, although not without heavy losses: a third of the colony may perish at this time.

After a brief rest in the caves of the Pyrenees, the mouse-eared bats press on towards Spain. According to Norbert Casteret, the bats cross Spain, and, after calling in at the caves on the Rock of Gibraltar, they continue on their way to their ultimate goal: by way of the littoral caves of Morocco (the Caves of Hercules) they finally reach the caves of the Atlas Mountains, where immense colonies live (up to 100,000 of these bats have been counted). These caves appear to be a focal point on which bats from France, Italy and North Africa all converge. It is here, at the end of August, that the females find the sedentary males. As autumn approaches the bats prepare to hibernate, having stored abundant fat in their tissues to last them during the long sleep. Copulation takes place just before their long sleep begins.

Shortly after the spring awakening the females congregate in distinct groups according to the country of their origin, and then they fly off, abandoning the males whose usefulness ended at copulation. The Mediterranean is recrossed, sometimes at its widest points, and the bats repair to their summer quarters.

The way in which bats navigate is astonishing. As is now well known, they are capable of flying in pitch darkness, avoiding obstacles by means of echo-location. For this they emit high-pitched squeaks in short bursts with frequencies of 30–100 kilocycles a second, the sound being emitted from the mouth except in the case of the Horseshoe bats (*Rhinolophidae*) in which the sound is sent out through the horseshoe-shaped fold round the nose. With their highly sensitive ears they pick up the minute echoes of these sounds as they bounce off objects, and from an almost instantaneous analysis of the time taken by the echo to return they can calculate the distance and direction of objects in their vicinity. If a container is divided into two halves by a mesh partition in which a small hole has been cut, a bat placed in it can pass from one side of the box to the other through the hole without hesitation, though the box is blacked out. The precision of this 'auditory vision' is even more extraordinary since fine wires of a tenth to a hundredth of an inch thick can be recognized as obstacles from three to nine feet away.

There are, however, occasions when the bats do not use their echo-location system, as for example when they fly in a place that is well known to them. Here they show an equally remarkable ability to remember their surroundings. Thus, they can navigate the often winding passages that lead from the entrance of their cave to the interior without relying either on sight, since it is dark, or on their echo-location, since a board placed across the passage will be crashed into, a thing that never happens in unfamiliar places. The discovery of these amazing navigational aids explains much but by no means all. Nothing is known, for example, about the means whereby bats are able to find their way over long distances. Some experiments by Norbert Casteret demonstrate this aspect of bat navigation.

One experiment consisted in catching bats (mouse-eared bats in this case) in a bag, placing them in blacked-out containers and releasing them from places that they had never visited and that were some distance from their caves. The results of this experiment were quite remarkable. Casteret caught the bats in the Tignahustes cave and took them by train to their destination. But, as the train passed the cave, the bats, confined in their light-free boxes, immediately started to show signs of nervous activity and set up desperate squeaks. How possibly could they sense the proximity of their home cave? This still remains a mystery.

The results of a second experiment were equally surprising. Bats were released some hundred or more miles from their home cave in a place totally unknown to them and well off their usual migration route. By marking the bats and searching the original colony it was found that they had managed to return very quickly. One female, for example, caught and marked on 15 May 1925 and released from Saint-Jean-de-Luz, was recaptured on 22 May at Tignahustes, its cave of origin, 125 miles away. This same female, released from Orleans, seems to have found its way back. It was found dead thirteen years later in a cave at Cigalière.

Much more work is needed before the migratory behaviour of bats is properly understood, but recent experiments on birds (see page 125 et seq.) may throw light on it.

Watching a seal at a zoo, dejectedly stretched out on a rock or nonchalantly floating with its nostrils just above the surface, it is hard to realize that these animals travel about 6,000 miles every year in the wild. But once the seal starts swimming one can see how this is possible, for with almost effortless movements seals can maintain average speeds of twelve knots (about fifteen miles an hour, or faster than a powerful yacht). The most characteristic of the migratory seals is the Northern Fur Seal of the Pacific (it should be noted, in passing, that these are loosely called 'seals' although strictly speaking they are eared-seals or otaries; true seals, such as the common and grey seals of British waters, have no external ears or pinnae and their fore-legs merely act as rudders and cannot be used for walking on land).

The males of the Northern or Pribilof Fur Seal (*Callorhinus ursinus*) are large creatures weighing over 600 lb., five times the size of the more graceful females, and they have a very im-perious air about them. The males, leaving the waters of Japan, take to the open Pacific and about mid-April turn north-east-wards and head for their rookeries on the islands of the North Pacific. By May males are arriving from all parts of the Pacific. Fighting with great ferocity for a strip of beach, they establish territories on these islands, which for the rest of the time are colonized by millions of true seals. For three months these power-ful ten-year-olds ceaselessly defend their territories without taking any food at all (to leave one's territory would be fatal). In about mid-June the females arrive and, having chosen a territory and its proprietor (or having been chosen), they give birth to their young within a day or two of their arrival. The females have, in fact, a very good appreciation of the progress of their pregnancy, for the end of the migration coincides exactly with the arrival of the single pup. After the young are born there is still some dispute on the question of the composition of individual harems, but these squabbles do not have the violence of the earlier battles for territories. Some days later comes the arrival of the two- to five-year-olds and, not having a territory to go to, they are relegated to areas further from the shore, where they alternate between play and sleep. The females become pregnant but will not have to care for their young until a year later, at the time of the next spring migration.

In mid-August the males, who have by now fulfilled their paternal duties, finally abandon the harems and swim to their fishing-grounds where they take some well-earned nourishment. From now on they will not use their territories again until the next season. It is at this time that the one-year-old pups and the virgin females of two years old appear on the scene, for the migration is not yet finished. At last the males who have been entirely celibate in the previous period, can set about selecting a mate. Towards the end of October, when the never very agreeable weather becomes really bad, the mass exodus to the south begins. Cold weather and daily storms force tens of thousands of fur seals southwards and this continues for two months until, by December, even the oldest and strongest males have fled. Passing southwards the migration splits, one lot heading obliquely eastwards towards Japan and the other pressing on southwards to winter off Mexico, especially off the island of Guadeloupe. It is along the American Pacific coasts that the northern fur seals most closely approach their southern counterparts, the Southern Fur Seals, which reach as far northwards as the Galapagos Islands. The southern fur seals (several species of *Arctocephalus*) have a rather similar migratory pattern.

Nearer home, the Grey Seal (*Halichoerus grypus* – a true seal) also moves north and south with the seasons, the old bulls arriving ahead of the females and establishing territories. Nowadays the grey seal penetrates as far south as the coasts of Cornwall although formerly they were found right down to Rugen. One population, the harp or Greenland seal from the central Arctic, makes for the Jan Mayen Sea between Spitzbergen and Iceland while another, from the eastern Arctic seas, congregates in the White Sea. The Common Seal (*Phoca vitulina*), the other seal commonly seen in British waters, reaches much further south, extending down as far as France and even Portugal.

Other seals, such as the Hooded Seal (*Cystophora cristata* – pouch in the snout, prefer the higher latitudes and in springtime breed on the drifting icebergs, although they too move southwards in winter.

We can end this section on migratory mammals with a consideration of whales, of which the Blue Whale (*Balaenoptera musculus*) is the largest of all living animals, reaching a length of some 90–100 feet. Whales are essentially inhabitants of the cold seas. True, they are found in the warmer oceans during the winter, but they are by no means widespread in tropical waters. This is principally because the tropical seas are virtual 'marine deserts'. The huge concentrations of plankton that abound in the temperate and cold seas do not occur in the tropics; the concentration of planktonic or free-floating micro-organisms is about 100 times greater in the Antarctic than it is in equatorial waters. There are, however, special areas, such as the Red Sea or the waters off the Galapagos Islands, where an unusually rich plankton coincides with the presence of whales.

It is also noticeable that the whales of Arctic seas are much smaller than those of the Antarctic, even when the same species is compared. This too can be attributed to the fact that the plankton is richer in the south polar seas. Thus one finds a very strict correlation between the presence and concentration of plankton and that of the whales.

It is remarkable that some of the smallest organisms in the oceans form the food of the largest animals on earth. How is it that these monsters can feed on such small prey? In the first place one must make a distinction between the toothed whales (the *Odontoceti*) and the whale-bone whales (the *Mystaceti*). The former group, which contains the Killer Whale (*Orcinus orca*) as well as porpoises and dolphins, are equipped with teeth in the jaws, whereas in the second group the teeth are replaced by long, thin horny strips, the 'whale-bone' used to support corsets and stays in Victorian times. The rows of whale-bone fit closely together and act like a sieve. In feeding, these whales rise to the surface and, taking a prodigious gulp, draw in a huge quantity of water. They then close the mouth and force the water out through the sieve whose fine hairs retain the minute plankton (fish larvae and small crustaceans as well as microscopic animals). The movement of the plankton is dependent on currents, but when high winds produce large waves, as in the middle of the southern Pacific or Atlantic, the plankton rises to the surface attracting whales, which follow the storms for that reason. Around the continent of Antarctica there are five major whale zones (see map on page 37) of which the Weddell Sea is the most important, for here there are several ocean currents that effectively concentrate the 'krill' (krill is the standard name for the plankton on which the whales feed – usually made up of small crustaceans known as euphausids). The following pattern of migration occurs when the hard winter sets in and the whales cannot approach the coasts because of the ice-floes, they emigrate to tropical waters and remain there for some months barely feeding at all (the stomachs of whales caught in the tropics are practically empty).

How have these migrations been determined? Our knowledge of whale movements came first of all from the accounts of whalers which provide the main outline of the pattern. After that an analysis of the stomach contents and the marine organisms that settle on the skin of whales showed that whales caught in the Antarctic had passed through tropical waters and those caught in the tropics had passed through the Antarctic. Again, the identification of pieces of harpoon left in wounded animals provided the first true confirmation of a whale's particular movements. Thus, in the northern hemisphere, a whale was caught in Norwegian waters which had embedded in its flesh a harpoon of undoubted American origin. In another instance a tooth-powder tin containing a message was found in the stomach of a whale; the name and address was given of a man who had been lost overboard during the 1953–4 whaling season some 40° east of the spot. This proved that whales migrate between the Antarctic and Australia.

It was in about 1920 that a scheme was begun for marking these huge marine mammals. Small copper darts shot into the skin of the whales often caused infection, however, and frequently the darts would fall out. Since then stainless steel tubes have been used which are embedded into the muscles of the back by means of a special type of harpoon. On the tubes is engraved all the information and instructions necessary to ensure that details of the recapture of the whale are passed on if the animal is caught again. Of some 5,000 animals tagged, 370 have been recaptured (7 per cent), which compares quite well with the returns got from other animals, birds and fishes. From the results of this marking programme it has been possible to work out the movements of Humpback Whales (*Megaptera nodosa*), the Common Rorqual (*Balaenoptera physalus*) and the blue whale which, as far as the southern hemisphere is concerned, migrate back and forth between the tropics and the Antarctic.

The Humpback Whale, which whalers recognize by its characteristic short and broad 'spout' of water vapour, leaves its winter quarters of the Antarctic and heads for various areas (see map on page 37), returning to the Antarctic, often to the same place again, the following summer. This whale has curious habits.

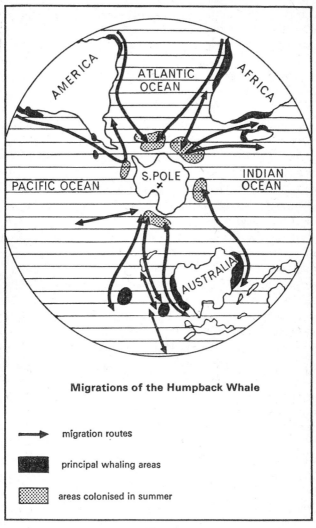

Migrations of the Humpback Whale

→ migration routes

▰ principal whaling areas

▨ areas colonised in summer

Fig. 4

Sometimes it leaps clear out of the water and then crashes down on its back with a tremendous splash; at other times it lies immobile on the surface. Humpback whales live in isolated colonies, which explains their faithfulness to their own Antarctic areas. This is not the case with blue whales and rorquals, whose migration routes are much less well defined. Thus, while a part of the rorqual population winters in the Antarctic, another part will make its way towards warmer waters, often at a considerable distance from the Antarctic. Rorquals even extend as far as north-western Africa, for individuals recorded in the winter from the Bay of Bengal, the Gulf of Aden and the Caribbean must surely be those from the Arctic, i.e. northern populations which, during the southern hemisphere winter, would be spending the summer in the Arctic. In spring the rorquals return to the Antarctic, where they arrive shortly after the blue whale. The pregnant rorquals, however, leave the Antarctic much later than the pregnant blue whales. It is interesting to note that temperature can affect the dispersal of these two species; for example, if the air temperature is below freezing point, the blue whales round the islands of South Georgia disperse before the rorqual, but the reverse is true if the temperature is above freezing point.

We can now turn to the northern hemisphere. In the species to be dealt with it is necessary to distinguish between northern and southern populations. While the pattern of their migrations is comparable (winter in the tropics, spring and summer in the polar seas) it is nevertheless reversed in time, so that the tropical waters are frequented by two populations of the same species but at different times of the year corresponding to the respective summers of the northern and southern hemispheres.

The behaviour of whales in the northern hemisphere is somewhat different from those of the south. The blue whale, for example, does not penetrate to such high latitudes as its relatives in the south, and the map of its migrations (see page 39) shows clearly that it tends to hug the shores.

The Greenland Right Whale (*Balaena mysticetus*), which before it was heavily exploited was so numerous that it was merely referred to as 'the whale', is typical of the northern seas. Although it does not leave the Arctic waters it nevertheless accomplishes a considerable journey throughout the year in its

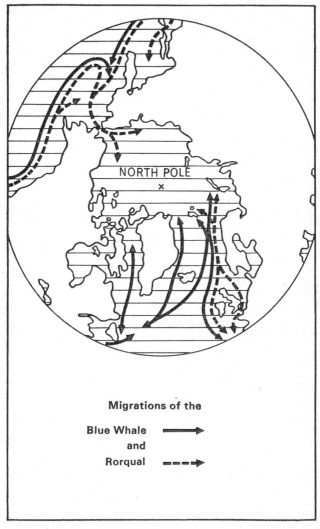

Migrations of the

Blue Whale ——→
and
Rorqual ----→

Fig. 5

search for rich plankton areas. The Black Right Whale (*Balaena glacialis*), before it became so rare, migrated as much in the northern as in the southern hemisphere; it was even seen off the coasts of France.

In the North Pacific one can find evidence of many migration routes. Thus populations of whales (blue, finback, rorquals, grey whales, humpbacks, etc) that have travelled along the Asiatic and North American coasts meet up before the Bering Straits, but on their return in the winter they revert to the same routes by which they came. Exchanges between the eastern and western populations occur but are fairly rare. One proof that the eastern and western populations remain fairly distinct was furnished by Dr Kazuo Fujino, who recorded the relative frequencies of different blood groups in the two populations. It is now well known that the cells of animals, and especially the red blood-cells, contain chemicals that are often quite specific for the species or even the population from which the animal came, and these can be correlated with the different blood groups. (The statistical study of the distribution of the various blood groups in man provides a precise and very elegant method in ethnological studies – blood group O, for example, occurs in 40 per cent of the population of Western Europe, but in 90 per cent in certain Indian groups.)

The Grey Whale of California (*Escrichtius gibbosus*) is particularly fond of coastal waters and does not hesitate to enter estuaries at the considerable risk of being stranded. The females have their young in the Californian lagoons between November and May. One curious habit in this whale which has often been observed is the way in which it will rise to the surface to breathe and place its head on a large block of ice, using it as a pillow.

To turn to the Sperm Whale (*Physeter catodon*), this species is toothed, and in former times provided a much sought after source of ivory. The cachalot or sperm whale also produced the precious ambergris used in the manufacture of perfumes. This large whale is very widespread in its distribution but it particularly frequents the tropical and temperate waters lying between 40°N and 40°S. From time to time it moves further north or south. Quite often males and females, young and old, move in distinct groups, following very well-defined migratory routes which the French whalers know and refer to as veins (i.e. areas of good luck).

Amongst the different species of dolphins and porpoises, which have a vast geographical distribution, there must certainly be some that are migratory, but as yet little is known about their migrations. In general, the movements of these remarkably intelligent mammals are mainly linked to those of their prey (herrings, etc.)

Other relatives of the whales remain within the Arctic Circle. This is the case for the Beluga (*Delphinapterus leucas*) and also the Narwhal (*Monodon monoceros*) whose curiously spiralled tusk (one origin of the unicorn legend) is found only in the males.

Birds

Turning now to birds, we enter the realms of the great migrants, whose movements have been known from earliest times and which provide some of the most striking and impressive examples of migrations over long distances.

There are a number of methods used to determine the migratory routes of birds and to plot these on a map. The oldest and quite reliable method is simply observation with the naked eye or with optical instruments, which enables one to say that a certain species was present at a certain time and place. It also can give an indication of the number of birds perched or flying past, their behaviour and, if the observations are repeated regularly, one can deduce the general pattern of the migration and the particular place that it occupies in the life cycle of the bird.

For this kind of observation it is essential to record the procedure (times and places accurately recorded, weather conditions, etc) in order to eliminate any bias due to the observer himself. Such observations can be made in the daytime and then supplemented by observations made at night on the basis of the sounds made by the migrants. Although less precise, and demanding more skill, this kind of record can nevertheless give useful information on the pattern of migration. It should be added that the recognition of bird song at night requires some skill since the calls of nocturnal migrants are often different from the calls they make in the daytime. Observations against the disc of the moon, on the other hand, provide mainly a quantitative estimate of the number of birds, although an experienced observer can certainly recognize the family to which the birds belong. Such observations are made with field-glasses or a telescope trained on to the moon. Birds crossing the line of sight show up very clearly and in spite of the very small segment of the sky visible the results are extremely useful. At the height of a migration it is not unusual to see birds outlined against the moon a dozen or so times in the space of fifteen minutes. The direction and eventually the duration of the flight can be noted. With a lens bearing a graduated scale it is even possible to judge the distance of the birds and thus estimate the speed of their flight. The limitations of this type of night observation are that they demand a clear night and an almost full moon (two or three nights either side of the full moon at most).

In recent years, radar has been found to be a useful tool in observing bird migrations. Radar provides a gross and purely quantitative estimate of movements within a range of about five to ten miles. A flock of birds is projected as a kind of cloud on to the screen – referred to as 'angels' until the cause was discovered.

Certain places offer special opportunities for watching bird migrations. Lighthouses are an example, for the light is a strong attraction to birds and they will often wheel round and round the light and sometimes throw themselves against it. There are, in fact, some obstacles that wreak havoc among migrating birds, one of the best known being a certain 100-ft television tower in France round which the bodies of 20,000 birds were found on a single occasion.

Finally, one of the most important supplementary methods for deriving data on bird migrations is ringing or banding. The birds are usually caught with nets of various sizes (most often a net thirty feet long and ten feet high) stretched between two poles. The nets are equipped with tightly stretched horizontal wires about two feet apart with the net loose in between and forming a series of small pockets. Birds flying into the net, which is made of black thread and is almost invisible against a green background, slide into the pockets and become entangled. The net is used to capture the birds at regular intervals and they are then ringed. Ringing consists of fitting an aluminium ring around the right foot of the bird. Inscribed on the ring is a serial number and the name of the organization making the study. The birds are then weighed and measured (wingspan, beak length, length of the tarsus, length of tail, and so on). The ring is generally fixed round the tarsus, i.e. normally the long part of the leg, except in swifts and martins where this part is short and the ring is then fixed to the angle of the wing. Information from the ringing is placed on cards in a central office and notification is sent to those who made the ringing if a recapture is recorded.

Ringing can also be done from the nest and this provides a certain amount of useful information since it shows the movements of the birds from the very start of their lives. It presents grave dangers for the fledglings, however, not only from the risk of damage due to handling or falling out of the nest in panic but also from expulsion by the parents, since many birds will not tolerate something strange or new in the nest, such as a foot with a ring on it would appear to be.

Apart from these individual methods of capturing birds, the large biological stations have larger and more sophisticated means at their disposal. Two can be mentioned here.

Heligoland trap: This is a vast curved funnel of wire mesh ending in a chamber with a non-return entrance: it is used for water birds, being generally placed on an open stretch of ground, into which the birds are driven.

Rocket nets: These nets are initially folded lengthwise and then, by one means or another, they are shot out over a flock of birds on the ground. If the net is very large and small rockets are used to project the two outer corners (detonated from some distance away), a large area can be covered. This is especially useful for ducks, geese and birds of the seashore. Although expensive, this method can result in the capture of many thousands of birds at one time.

The number of birds that eventually provide information (i.e. ones that are recaptured by someone who is ringing, or that are found dead and the rings returned) is extremely small – about 2 per cent for swallows, 0·7 per cent for linnets, 1·1 per cent for finches. The percentage is considerably higher for birds that are hunted for sport, such as the greylag (24%), shoveller duck (20%) and teal (16%), and for birds of prey such as the sparrow-hawk (15%), barn-owl (13%), and kestrel (12%).

The enormous numbers of birds ringed nowadays (600,000 annually in the United States, 120,000 in the Soviet Union, for example) require the most up-to-date methods of filing this information, sometimes by computers, but it gives an extremely accurate indication of the routes followed by a number of migratory species. Nevertheless, ringing is not without snags, centring round the effect of the ring on the behaviour of the bird (the capture itself is a traumatic experience while the shining ring may have unpredictable effects on breeding behaviour, courtship display, selection of mates, and so on). There is now a saying that 'we understand the migrations of *ringed* birds – but what about the others?'

Before dealing with the orientation of migrating birds, how they find their way and reach a goal they have never before seen, and before discussing the physiological changes that occur before and during migration, we should first review the occurrence of migratory birds in different parts of the globe, with particular reference to those that are seen in Europe. Without attempting to cover all migrant species, which would be impossible in a small book of this sort, we shall nevertheless try to give as complete a picture as possible so that those who want to know about the journeys made by a particular species will find at least some mention of them here; for this reason, western Europe is dealt with more fully than are other sections.

1 Europe

First of all some general considerations. The migrant birds of Europe often vary in the degree of their migration. Thus it has been found that birds that nest in northern and eastern Europe are more strictly migratory than those that nest in western Europe. This phenomenon, which only affects 'partial migrants' (for the long-distance migrants are fully migratory), is almost certainly due to climatic factors.

On the other hand, partial migrants in islands (especially the British Isles) often have a semi-sedentary behaviour. As distinct from those, movements are also found among the young of many species that are quite distinct from true migrations and which may be termed 'post-juvenile dispersions'. This occurs, for example, in the young of the tawny owl (of which the adults are sedentary), and the starling: in June and July one often sees whole flocks composed entirely of juveniles, recognizable by their plumage, which is a lighter grey and more uniform than that of the adult. Such birds disperse on leaving the nest, going in all directions and avoiding heavy concentrations of individuals round the nesting sites. This radial dispersal has neither the same motivation nor the same characteristics of a true migration and should not be termed such.

In Europe there are two main axes of migration, the first north-east by south-west and the second north-west by south-east.

According to recent statistics, of the roughly 450 species of birds in Europe, 175 are migratory; in fact, if one takes into account the partial migrants, it leaves only about 60 species that are completely sedentary. Of the true migrants, 150 follow the first axis of migration, 35 of them going as far as tropical Africa. The other 25 follow the second axis of migration, 14 of them reaching East Africa and 3 the Far East.

What are the main features of the journey to Africa? The first obstacle is the Mediterranean, which some avoid by crossing at Gibraltar or the Bosphorus, while others make the full journey from one side to the other. Migrants that make for North Africa or the Middle East have no more hurdles to cross, but for those that press on further to the south there is the vast expanse of the Sahara Desert to contend with. Many attempt to cross it but success is by no means guaranteed, and numbers of birds fail. Some species are apparently obliged to make the journey in a single step (about sixty flying hours), but it seems that many find rest at an oasis. Another set of migrants takes the parallel route along the Nile valley and reaches Africa with less difficulty. It is, in fact, the north-east or Ethiopian region of Africa that is richest in migrants from Europe. In general, our European migrants colonize a very varied range of habitats when they reach Africa, from the scrublands of the southern Sahara to the swamps of Lake Chad, only the thick equatorial forest being avoided.

In the following review the systematic order followed is approximately that adopted in the *Field Guide to the Birds of Britain and Europe* (except that sea birds are placed in a special section). That book is, incidentally, invaluable to those who want to identify the birds mentioned here.

Divers

The diver is a symbol of untamed nature. It comes from the far north and is very shy (although I have been able to approach within a few yards of it in a catamaran) but is one of the most skilful of fishermen. It remains under water for up to a minute and a half, and dives to a depth of thirty feet in its search for food in the shape of small fish. Its body is, naturally, elongated and its beak serrated. The Great Northern Diver (*Gavia immer*), of which the male assumes a black and white chequer-board plumage in spring, nests in North America, Iceland, Jan Mayen Island and Greenland. In winter its southerly migration regularly touches our coasts. The Red-throated Diver (*Gavia stellata*) nests in the more northerly parts of Europe, Asia and America and all around the North Pole. The 'great white silence' of the Arctic is filled with its wild cries and plaintive calls. In winter it comes southwards and is distributed along the coasts from the North Cape to Gibraltar and also on certain inland lakes.

Grebes

Grebes are exclusively aquatic birds that do not take readily to the air and are even more reluctant to venture on to dry land. The best known is the Great Crested Grebe (*Podiceps cristatus*), which is the more common of the two. It is quite easily recognized by its long neck and the double tuft of grey feathers on its head. In springtime it has further tufts of reddish and black feathers on the sides of its head. It is a partial migrant and the British population often moves southwards in winter to augment stationary populations in Europe. The Little Grebe or Dabchick (*Podiceps ruficollis*) is a roundish, tail-less bird found in still water, and it too is a partial migrant, sometimes wintering in western or southern Europe.

Herons

The migratory behaviour of European herons varies enormously, not only in different species but also from place to place and even between individuals. From the sedentary (e.g. in Great Britain) to the fully migratory, one finds all intermediates so that the herons are excellent examples of partial migrants. In the autumn the Grey Heron (*Ardea cinerea*) tends to move towards the south-west, so that those from northern Europe migrate to France while those from central Europe reach Africa.

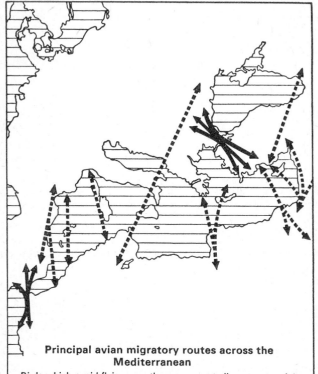

Principal avian migratory routes across the Mediterranean

Birds which avoid flying over the open sea at all costs are mainly those that are large, fly by day and use a gliding type of flight (notably storks and birds of prey). Their routes (solid lines on map) are thus restricted to the Gibraltar crossing (aligned generally with the SW—NE migration direction) and the Bosphorus crossing (almost at right angles to their migration route).

Other migrants, the majority in fact, brave the sea crossing, generally at night, which makes recording of their flights more difficult.

Certain routes are clearly more favourable than others in that they reduce the time actually spent over the sea — always a time of danger for birds. Routes that are at least partly over the sea are shown by dotted lines.

Fig. 6

In France, which benefits from an equable oceanic climate (in comparison with Scandinavia, for example), the herons are less inclined to emigrate. Many remain where they are, but some do migrate, a few to the north but most to the south-west. The more adventurous, however, make their way across the Atlantic, for quite often birds which have been ringed in France are recaptured in mid-Atlantic. This feat is better understood if one considers the flight of the heron. It has a long and slow wing-beat, a steady, regular flapping with the neck drawn back and the feet stretched out behind, and it seems to be perfectly confident over the ocean. Birds with a gliding type of flight are much more awkward over the sea where the winds, aerial turbulence and up-draughts are quite different from those encountered over the land. The other members of the *Ardeidae* or heron family are more markedly migratory than the grey heron. The Purple Heron (*Ardea purpurea*), Egrets (*Egretta* species), the Bittern (*Botaurus stellaris*), the Squacco Heron (*Ardeola ralloides*) and the Night Heron (*Nycticorax nycticorax*) all spend the winter on one side or the other of the Mediterranean basin, often reaching tropical Africa.

Storks

The White Stork (*Ciconia ciconia*) is a very well-known migrant with as firm a place in legend and fable as the swallow or the turtle-dove. This splendid bird, that has for centuries nested on the roofs of houses in various parts of Europe, appears each year and heralds spring. Quite widely distributed in Europe, it spends the winter in Africa. It avoids a direct passage over the Mediterranean since it is better suited to flying over the land. To do this, the European population splits into two groups, the first passing through France, Spain, over Gibraltar, across Morocco and into central Africa, the second migrating south-east and crossing the Bosphorus before passing down into Africa, sometimes penetrating as far as South Africa. In Africa the storks live a vagrant life and there are exchanges between those from the eastern and those from the western migration routes. Two points in the journey, Gibraltar and the Bosphorus, act as funnels through which the birds must pass. Hundreds of thousands of storks pass over Gibraltar, but almost double their number take

Migration routes of the European White Stork

(After Rüppell, modified by Verheyen, 1950)

Fig. 7

the eastern route and an extraordinarily impressive sight these migrations are (one can see as many as 50,000 storks in the Gulf of Suez flying just above the water in a stream thirty yards wide and up to twenty-five miles long).

The return journey is the reverse of the outward migration. Storks sometimes return with curious souvenirs of their travels. There have been many cases of the birds returning with pieces of arrow embedded in them. The distinctive construction of the arrows and the marks carved on them are enough to identify the tribe and one can then pin-point the winter quarters of the unfortunate bird. The beautiful but much rarer Black Stork (*Ciconia nigra*) makes a similar migration but nests in the marshy forests of Poland, Roumania and Russia.

Spoonbills and Ibises

The Spoonbill (*Platalea leucorodia*) is another beautiful bird. It has spotless white plumage with a little tassel of feathers behind the head, but its most distinctive feature is the flattened, spoon-like tip to its bill. Spoonbills formerly nested in France, but they still nest in Holland, Spain (Guadalquivir) and the marshes of south-eastern Europe (Montenegro, Bulgaria and Roumania as well as the sanctuary of Neusiedler Lake on the border of Austria and Hungary). Spoonbills return to Africa in winter and the map shows that the winter quarters of the different European populations are quite distinct.

There is one ibis that is found in Europe, namely the Glossy Ibis (*Plegadis falcinellus*) which returns from central and southern Africa in April and May and nests in Guadalquivir and south-east Europe, where it is quite common.

Anatidae

This large family comprises the swans, geese, ducks and mergansers. It thus includes numerous web-footed species, amongst which are some very interesting migrants. In addition to the Mute Swan (*Cygnus olor*), there are two other species in Europe, the Whooper Swan (*Cygnus cygnus*) and Bewick's Swan (*Cygnus bewickii*). The last two are northern birds, especially Bewick's swan which nests along the coasts of Siberia and Novaya Zemla.

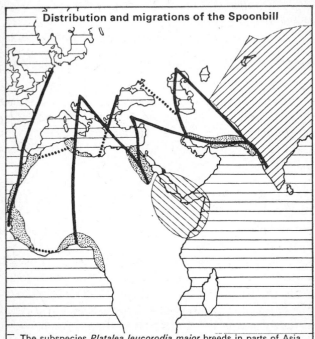

Distribution and migrations of the Spoonbill

The subspecies *Platalea leucorodia major* breeds in parts of Asia (upper hatched area on map). The subspecies *P. leucorodia leucorodia* breeds in colonies extending as far as the Atlantic, i.e. – the Arguin sand-bank (Mauritania), 2,000 pairs – Low Countries, Turkey, Volga delta, about 500 pairs – Andalusia, Danube delta, Crimea, Sea of Asov, 100 pairs. Migration routes are shown by lines (dotted where still hypothetical) and the wintering areas are stippled. The segregation of the populations and their double migration routes are immediately noticeable, one to the east and one to the west. A mutation – absence of the yellow cross at the end of the beak – appeared in the Arguin colony and was also found in the colonies of Andalusia and Roumania, but not in the Low Countries (Vieillard, 1962). This unusual but highly significant fact can be taken to show that there is a connection between these areas, although much more investigation is needed. The Ethiopian subspecies *P. leucorodia archeri* is indicated by the lower hatched area.

Fig. 8

In winter this species flies south-west and spreads out across the Baltic and the North Sea and in a particularly hard winter it may come as far as France. The whooper swan reaches even further south.

Wild geese – the phrase has a magical ring about it, particularly for naturalists. Coming down from the north, they seem to bring with them a breath of the wilds, of mysterious and solitary places. It is always a thrill, on a cold December morning, to hear the cry of geese breaking through the misty stillness. It may be a White-fronted Goose (*Anser albifrons*) arriving from the coldest parts of the globe (Siberia, Greenland, Russia) and perhaps deciding to stay here a while. Again, it might be the Greylag Goose (*Anser anser*), the most southern of all the geese since it not only nests in Russia, Scandinavia and Scotland but also in Bulgaria and Roumania. It can be identified by the strikingly pale leading edge of the wing and by the more sombre colour of the body. It may also be the Pink-footed Goose (*Anser brachyrhynchus*) on its way from Greenland, Spitzbergen or Iceland for the winter. Also seen in this country is the Bean Goose (*Anser fabalis*) which, escaping the rigours of the Arctic where it breeds (Scandinavia, Finland, Russia and eastern Siberia), finds refuge from the cold, but not the sportsman, in the Baltic and as far as southern Europe.

Turning now to ducks, their migrations have certain peculiarities. The first is the 'moult migration'. Most birds, when they moult, lose their feathers one by one so that when the last have fallen the first have regrown. This is not, however, the case with ducks, for after nesting they lose all their long wing feathers (primaries) at once and are then incapable of flight until the feathers have grown again. Obviously birds such as ducks, which are wild and heavily preyed on by man and other predators, must have to adjust their migratory behaviour in the face of this peculiarity. For this reason there is a special type of migration between the time of mating and the time of the autumn migration and this is one that takes the ducks to some safe place where they can moult in peace. Usually this is an area with swamps or marshland where they can hide amongst the vegetation. The second part of the migration takes them to their winter quarters and the last part leads them back to the nesting grounds.

It is in the latter part of the migration that the second peculiarity occurs. This involves a certain inconsistency and unfaithfulness

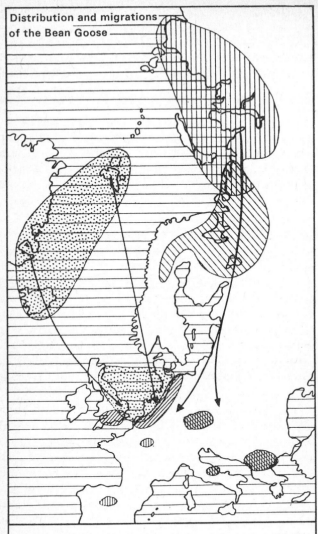

Distribution and migrations of the Bean Goose

Three subspecies of this goose winter in Europe, i.e. *Anser fabalis brachyrhynchus* (stippled), *Anser fabalis fabalis* (oblique hatching), *Anser fabalis rossicus* (vertical hatching). Their wintering areas are quite localised, although the species may wander across the whole of Europe. The subspecies *brachyrhynchus* occurs in a restricted area, with rather little contact with the other two. The latter, however, mix to some extent in their nesting areas and disperse haphazardly to favourable areas throughout Europe.

Fig. 9

to their nesting areas. Most migratory birds return to nest year after year in perhaps not exactly the same place but at least the same area. In ducks, however, there is a great variation. One year they will nest in Great Britain, the next in Scandinavia, or they might equally well turn up in Russia or Scotland. There seem to be several factors involved. In the first place ducks, which come into breeding plumage very early, often mate in winter, and the fact that birds from different populations are mixed together in the wintering area may mean that one or other partner may be led back to an alien nesting area. It is also thought that a duck, mixed in with birds from other regions during the migration, might be led along a different migration route than the one that it arrived by. Sir Landsborough Thomson has termed this phenomenon *abmigration*.

The route followed by migrating ducks is closely linked to features in the terrain. Ponds, lakes and water-courses serve as landmarks and from a conservation point of view it is important that these should be recongized as such. For example, some key point may be eradicated in part of a development project (a pond drained for road construction, for example) and this can have a serious effect on the migration of ducks over this area.

Numbers of ducks winter as far away as Africa. The Gadwall (*Anas strepera*), which nests in various localities throughout Europe, spends the winter in the Mediterranean basin and as far away as Nigeria.

The majority of whistling teal that breed in Europe (British Isles, Scandinavia, south-eastern Europe) spend the winter along the coasts of France, but others continue southwards and congregate in tropical Africa. The Pintail (*Anas acuta*) reaches as far as that, as also the Shoveller (*Anas clypeata*) although most of them do not pass the shores of the Mediterranean or North Africa.

The Mallard (*Anas platyrhynchos*) is a partial migrant and the sedentary populations are augmented in winter by the great flights from the north-east of Europe. England also receives a contingent of Pochards (*Aythya ferina*) coming from central Europe or even the Urals.

The Tufted Duck (*Aythya fuligula*) does the same and then continues further south towards Liberia and Kenya.

The Velvet Scoter (*Melanitta fusca*), which can dive down to a depth of sixty feet and remain below the surface for three

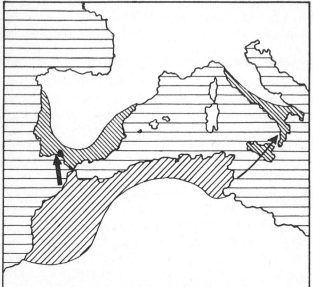

Migrations of the Ruddy Shelduck

This Asiatic duck has an independent population in the southwest of the palaearctic region, and its normal area of distribution is shown in black on the map. The main nesting area is in the Middle Atlas region of Morocco and the nearby desert and coastal region. After breeding, a part of this population makes for the Marismas of Guadalquivir in Andalusia. These ducks spend the autumn there and then return to their country of origin, Morocco, during the winter or at the beginning of spring. This route is shown by a thick arrow to the west of the Straits of Gibraltar. This is the only case known, discovered recently by Valverde (1960) and still not properly understood, of a species that breeds in Africa and winters in Europe. Such a movement can be likened to the phenomenon of the northward spread in summer of certain Saharan birds, but a sound hypothesis to account for the movements of the Shelduck has still to be formulated (Vieillard, 1966). The hatched areas show accidental appearances of the Shelduck. These may represent individuals that have overshot their winter area at Guadalquivir or those that have taken the route Cap Bon – Sicily – Italian Peninsula (thin arrow).

Fig. 10

minutes, migrates from Scandinavia, Novaya Zemla and Russia to hibernate from the Baltic to the Gulf of Gascoigne.

The Common Scoter (*Melanitta nigra*), a more exclusively maritime species, sometimes reaches the coasts of Morocco.

The mergansers are partial migrants, the Red-breasted Merganser (*Mergus serrator*) leaving northern riecounts in Europe and heading for the coasts of France.

The Garganey (*Anas querquedula*), known as the 'summer teal' in France, is a summer visitor which migrates back to Africa in the winter. The migrations of Teal (*Anas crecca*) are very well known since it has been the object of intensive ringing operations and its great popularity with sportsmen has meant that about one in five ringed birds are subsequently traced. (In the Camargue, radiographs of the ducks have shown that the majority of birds that have been ringed have some shot in them, either from being hit by stray pellets or from swallowing shot while grubbing on the bottom for food). Their migrations regularly follow a north-east/south-west axis.

Birds of prey

The flying abilities of birds of prey are very varied and one finds all sorts of intermediates between the gliding that eagles and vultures can maintain for tens of miles and the lightning dive of the peregrine falcon.

Among the eagles one finds many migrants: the Spotted Eagle (*Aquila clanga*) is a migrant, spending the warm season in the north-east of Europe and the winter in Asia Minor and the Indies. The Golden Eagle (*Aquila chrysaetos*) is the only eagle regularly seen in Great Britain, and is sedentary.

The Lesser-spotted Eagle (*Aquila pomarina*), which nests in eastern Europe (Russia, Roumania), leaves in September for South Africa; these large birds seem to have no hesitation in making this immense journey.

Once widespread in France but now rather rare, the Booted Eagle (*Hieraetus pennatus*) retreats to the southern half of Africa while the populations from eastern Europe make for India.

The Buzzard (*Butea buteo*) has quite a complex migration pattern and one can distinguish a number of different populations with different migratory behaviour. Some are sedentary while others migrate in flocks and can move from Scandinavia as far afield as the southern parts of Asia and Africa. One Scandinavian subspecies or race comes down to France and joins the sedentary populations there.

The Rough-legged Buzzard (*Buteo lagopus*), which nests in the tundra areas of Russia and Sweden, takes flight for the south-west in autumn and heads towards England and Central Europe.

The Sparrow Hawk (*Accipiter nisus*) and the Goshawk (*Accipiter gentilis*) are partial migrants or sedentary.

Of the European kites, the Black Kite (*Milvus migrans*) is the most strongly migratory. Early in the summer it quits northern Europe and heads for central and southern Africa. The migrations of these kites often bring a very great number of these birds together and such migrations are particularly spectacular in places like the Bosphorus where there is a migratory 'funnel' (see page 48 and the discussion on the reluctance of birds of prey to cross large stretches of water).

With experience one can distinguish the Honey Buzzard (*Pernis apivorus*) by its longer tail, wings that droop very slightly when gliding and the fact that it soars and hovers less than the buzzard. As its Latin name implies, it feeds principally on bees, although it varies its diet sometimes to include reptiles (which does not save it from persecution by preservers of game). At the end of the summer the honey buzzard leaves not only France but the whole of central Europe for the tropical forests of Africa (Cameroons, Congo and Angola). These migrations sometimes bring together thousands of birds and present a magnificent sight.

The harriers are agile and elegant birds, especially when they hunt close to the ground. The most common is the Marsh Harrier (*Circus aeruginosus*), which is a partial migrant. Like many other species, it is for the most part sedentary in France but migratory in northern Europe. In the course of these migrations it visits equatorial Africa. The same is true of the Hen-Harrier (*Circus cyaneus*) which besides this reaches also as far as the Indies and China.

The Short-toed Eagle (*Circaetus gallicus*) is a magnificent bird with huge orange eyes. It feeds almost exclusively on reptiles and is a regular victim of misinformed country people; this is the more tragic because it lays only a single egg. It inhabits the southern parts of Europe and North Africa, and in winter the

European population departs for the Niger basin, Togoland, Senegal and Ethiopia.

The Osprey (*Pandion haliaetus*) is another highly attractive bird. It can be recognized in flight by its dark upper parts and snow-white under parts, the latter bearing rows of small black dots; also, the wings are very distinctly angled. It can be seen over large lakes, rivers or along the sea shore falling vertically and feet first more than thirty feet, breaking the surface with a splash and emerging with a fish between its talons. The osprey, which is very widespread throughout the world, breeds especially in Europe. In autumn birds from northern Europe reach as far as southern and tropical Africa in search of unfrozen water; quite at home over the sea, they often cross the Mediterranean at its widest point.

Falcons are noble birds with a long history of association with man. They have long pointed wings, long tails and the wing strokes are strong, rapid but shallow. The Peregrine (*Falco peregrinus*) is one of the most skilful hunters and can kill birds of all sizes in mid-air, often those that are larger than itself (herons, geese, ducks), dropping on to them at the amazing speed of over 100 miles an hour. If the peregrine does not succeed in seizing the bird, which is stunned if not killed by the blow, it will quite often retrieve it before it reaches the ground. The object of intense persecution throughout Europe, the peregrine now nests rather rarely in Europe and its preservation requires the utmost vigilance. The young peregrines from northern and eastern Europe and Russia spread to the south-west during winter and some of them go as far as tropical Africa. The Hobby (*Falco subbuteo*) looks rather like an overgrown swift, for it has narrow wings and a rather short tail; it does in fact prey on swifts, as well as flying insects at dusk. It is a regular migrant which reaches as far as South Africa when the insects on which it feeds disappear in our latitudes with the first frosts.

The commonest of the day-flying birds of prey in Europe is the Kestrel (*Falco tinnunculus*) and it is this bird that one sees making a protracted search for field-mice, hovering in one place for minutes on end, head to the wind, then slanting steeping down to pounce. It is ridiculous that these birds should be shot since they do a great deal to keep down unwanted rodents. Almost sedentary in England and France, the kestrel becomes more and more migratory towards the east and north of Europe, where it is a regular migrant to southern Europe and Africa (see map on page 61). On the return journey, kestrels have been ringed on their pre-nuptial migration at Cap Bon in Tunisia and recaptured as far away as Russia.

The Lesser Kestrel (*Falco neumanni*), of which some rare colonies are still found in the south of France, is more widespread in other Mediterranean countries. It migrates to South Africa.

The nocturnal birds of prey are for the most part sedentary, but there are some fast-flying migrants. It is, in fact, the smallest of them, the Scops Owl (*Otus scops*) – barely the size of a sparrow – which travels the furthest, travelling from southern Europe to spend the winter in equatorial Africa.

The Short-eared Owl (*Asio flammeus*) is a more irregular migrant but it traverses the whole of Europe, from Russia to France, and makes for Africa (Sudan, Kenya). The Snowy Owl (*Nyctea scandiaca*), pressed by famine in the years following the lemming migrations (see above, page 27), makes desperate journeys in search of food and these may take it far from Scandinavia. It has recently bred in the Shetland Islands.

To finish with this section, it should be said that, of all the animals that have fallen victims to progress and civilization, the birds of prey have come off perhaps the worst, and many species are now threatened with extinction (bearded vulture, sea eagle, golden eagle, peregrine, etc). This results from the destruction of their food supplies, coupled nowadays with the widespread and often irresponsible use of insecticides and other chemical products, which not only poison the prey species but also the birds themselves, leading to sterility, and sometimes to death. On the other hand, there have been encouraging recent signs of peregrines becoming acclimatized to urban habitats, largely free from poisonous food, and a correspondence in *The Times* in October 1968 showed their occasional fondness for tall buildings in London.

Gallinaceous birds

The only member of this group in our region that is recorded to have a migratory habit is the Quail (*Coturnix coturnix*). After hibernating in Africa, above or below the Sahara, it returns to

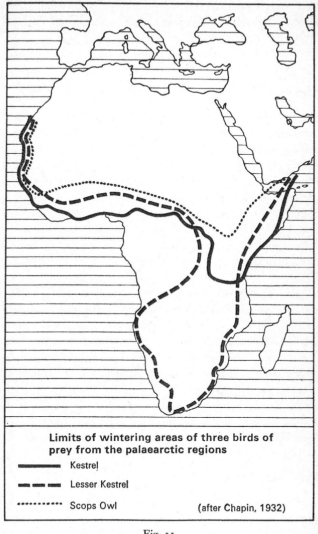

Limits of wintering areas of three birds of prey from the palaearctic regions

━━━━━━ Kestrel

━ ━ ━ ━ Lesser Kestrel

•••••••••• Scops Owl

(after Chapin, 1932)

Fig. 11

Europe, reaching Italy in March and France from April to May. For this journey it crosses the Mediterranean in a single night, flying only a few yards above the waves. During the day the migrants rest on various islands where they are found in very great numbers and at the same time each year. The quail only reaches the British Isles in very small numbers.

Cranes

One generally thinks of cranes as tropical birds, but this is a misconception. For example, there is the Crane (*Grus grus*), which has a wingspan of six feet and is the largest bird in Europe. It nests in the northern parts of Europe, chiefly in Germany and Russia. The crane is one of those species that migrate along a narrow route; the most characteristic of such narrow-route migrants are the storks (see page 49). After gathering in groups that may comprise thousands of cranes, the birds take off in autumn for their African territories. To get there they use narrow routes so that year after year they pass over the same spot following routes used by their remote ancestors. Three principal routes have been recognized, one of which crosses France (see Fig. 12). If one happens to be on this route (say in Champagne, Sologne or Basses-Pyrénées) one can hear the loud cries of the cranes and see their formations (a single line of birds in a V-shape) passing overhead at an altitude of up to 15,000 feet. Their supple movements, with strong 'rowing' wing-beats, enable them, like herons, to cross the Mediterranean or other stretches of water in a single night. Those that pass over France are heading for North Africa, and the flocks from Russia make their way to the Blue Nile and Somaliland.

Rallidae (Rails, Crakes and Coots)

With the exception of the Coot (*Fulica atra*) and the Moorhen (*Gallinula chloropus*), the birds in this family are rather shy, making observation difficult. With a little patience and with a hide built near some convenient gap in the reeds, one can see a Water Rail (*Rallus aquaticus*) or one of the crakes. The family includes some long-distance migrants, such as the Spotted Crake (*Porzana porzana*) which nests in an area from the Mediterran-

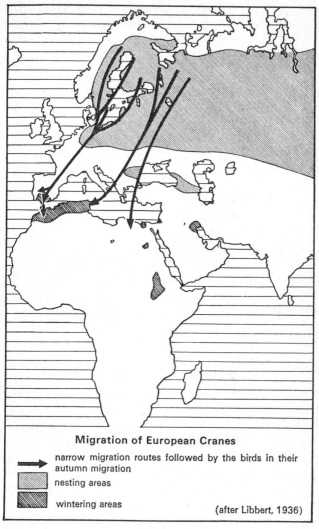

Migration of European Cranes

narrow migration routes followed by the birds in their autumn migration

nesting areas

wintering areas

(after Libbert, 1936)

Fig. 12

ean to Scandinavia but winters from North Africa across to the Sudan; or the Corncrake (*Crex crex*) which sometimes reaches as far as South Africa, but for the most part stops short in the equatorial regions. In April and May Europe is populated with the spring migrants. The corncrake is sometimes referred to in the south of France as the king of the quails, for a solitary specimen can be seen accompanying about a dozen quails on their migration, and standing out by reason of its size.

The coot, common on European lakes, is a nocturnal migrant capable of flying long distances at each stage of its journey (there is a case of a coot covering 450 miles in a day and a half). Coots from Germany, Poland, and even Russia winter in France and it is one of the species that is particularly faithful to its wintering site, returning every year to the same place.

Bustards

Becoming rather rare because of the attentions paid to it by sportsmen, the Great Bustard (*Otis tarda*) moves southwards from the Scandinavian countries to escape from the cold and snow. It joins sedentary populations in southern Europe. The Little Bustard (*Otis tetrax*) either winters in the very south of Europe or heads for Africa.

Limicoles

Members of the family *Scolopacidae* (sandpipers, godwits, curlews and snipe) are great travellers. One only has to watch the movements of a small covey of sandpipers to appreciate this. They peck nervously at the muddy sand-flats, run a few yards, take off suddenly with a series of shrill notes, fly rapidly, glide and settle once more to their feeding. It is not surprising that these birds are among the greatest migrants. But their extreme similarity makes the various sandpipers a nightmare to the budding ornithologist. With practice, however, one soon learns to distinguish them by their flight, cries and appearance.

The Sanderling (*Crocethia alba*), which runs like some mechanical toy across the sands, nests among the stony Arctic tundra. It occasionally visits Britain, and is common along French beaches. It winters as far away as South Africa, and other populations reach Patagonia (from the extreme north of America) as well as Australia. The journeys often start early and finish late. Many members of this group disperse and settle along the migration route so that the strongest fliers reach the most distant winter quarters.

The Curlew (*Numenius arquata*), for example, is found wintering from the north of France right down to the Cape of Good Hope, and it is only in this group that such vast wintering areas are found.

The Little Stint (*Calidris minuta*), scarcely larger than a robin, has a feverish kind of disposition but is very self-assured. Scarcely has it completed nesting in the extreme north (Siberia, Russia, Norway) than it hurries off southwards to the tip of southern Africa, where with the approach of spring it hastens back to our hemisphere with the same speed.

The plovers (family *Charadriidae*) are universally recognized as the boldest of migrants. Those that are found in Europe uphold this tradition well, and the Grey Plover (*Charadrius squatarola*) makes the return journey between the Siberian tundra, Alaska and the Cape of Good Hope. As in other limicoles, many birds end the migration early and the wintering area begins in the latitude of the British Isles.

Others that reach as far as South Africa are the Turnstone (*Arenaria interpres*), the Ruff (*Philomachus pugnax*), the Common Sandpiper (*Tringa hypoleucos*), the Greenshank (*Tringa nebularia*), the Ringed Plover (*Charadrius hiaticula*), and so on. Greenshanks have even been caught as far as the Kerguelen Islands, having doubtless been carried there from Africa and Madagascar by storms from the west.

There are others which do not reach so far south, of which one cane cite the Oystercatcher (*Haematopus ostralegus*), the Black-winged Stilt (*Himantopus himantopus*), the Little Ringed Plover (*Charadrius dubius*), the Dunlin (*Calidris alpina*), the Bar-tailed Godwit (*Limosa lapponica*) and the Black-tailed Godwit (*Limosa limosa*), the latter wintering particularly in the Lower Senegal valley, according to Francis Roux.

Although evidence has recently shown that limicoles undertake overland migrations, most of the species found in Europe follow the line of the coasts. But the habitats that suit them and provide both food and safety during their migrations are beginning to become rare, and this is particularly true with respect to their need for safety and peace. Such remaining areas must be carefully preserved, for projects like the drying-out of the Bay of Aiguillon would be a veritable catastrophe, since it forms one of their main landing-places.

Leaving aside the *Laridae* (gulls and terns), which will be dealt with separately under the heading of sea birds, and before passing on to the passerines, mention should be made of the pigeons and doves (family *Columbidae*). The migrations of pigeons are well known and the often massive movements of the Wood Pigeon (*Columba palumbus*) are familiar to the sportsman. The wood pigeon, however, is only a partial migrant. The migratory fraction of the population undertake regular journeys and readily cross the high mountains that they encounter on their way to Spain and North Africa. The Turtle-dove (*Streptopelia turtur*) is a most punctual migrant which, from the end of August, makes its way (mainly at night) towards the southern part of the Sahara region and particularly the Sudan.

Passerines

This is the largest of all the orders of birds. It would be impossible to list all the migrant species in the available space, but we shall try to review a fair number for the benefit of those who want at least some information of the group.

The Cuckoo

The call of the Cuckoo (*Cuculus canorus*) is quite unmistakable and its parasitic habit of laying its egg in the nest of another bird, even that of the minute wren, hardly needs mention. However seldom one actually sees a cuckoo, its familiar call regularly announces its arrival in April. But soon, by mid-July, the adults are off for the southern half of Africa, the young birds setting out to join them in mid-August, though occasional stragglers may be found in England after the beginning of September.

The Nightjar

This is a strange-looking bird. Seen close to, its curious head and eyes with a long pale 'eyebrow' above, together with its

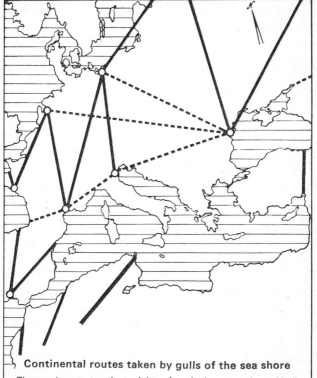

Continental routes taken by gulls of the sea shore

These migrants travel at night, often in large numbers. Until recently, the study of their migrations seemed to be the most difficult of all, but nowadays these birds have become choice research subjects in migration studies. Ethological-ecological studies (i.e. behaviour in relation to environment), such as those by Vieillard (1959), have led to more convincing interpretations of the movements of these birds. The need for stopovers (marked as circles here) between the breeding and wintering areas is now recognised, such places offering a reasonable food supply. Movements between these locations thus take place at night. Observations have enabled us to sketch in the axes of migration (solid lines) as well as less rigidly defined routes (broken lines).

Fig. 13

twilight habits, give it a rather mysterious air. In the gathering darkness of a summer evening an insistent and faraway humming noise reveals the presence of a Nightjar (*Caprimulgus europaeus*). If one catches sight of it, its silhouette reminds one of a large swift, while on the rare occasions that I have seen it by day I have thought it similar in appearance to a small falcon. After raising two broods the nightjar makes its way quickly to the south-eastern corner of Africa. It returns to Europe at the beginning of May.

The Swift

The Swift (*Apus apus*) has in common with swallows a graceful and thrilling style of flight, and is also capable of lightning speed. It should not, however, be forgotten that the swallow belongs to quite a different family, namely the *Hirundinidae*. The swift belongs to the *Apodidae*, or 'without legs', a reference to its short legs. Probably no bird spends less time on the ground than the swift. When night falls, the swifts rise, higher and higher in the sky, so that one wonders how they manage to sleep. The swift is also a good example of a punctual migrant. It seems to be equipped with a kind of internal clock (see page 116) for its arrival in spring recurs with astonishing regularity. It leaves early, from August onwards, in order to avoid even the earliest intimations of winter. After wintering from the Cape of Good Hope to West Africa, it returns to us quickly, having crossed desert and sea.

The Alpine Swift (*Apus melba*), which can be identified by its stronger flight, larger size and white front, makes the same journey but arrives sooner and leaves later.

Bee-eater and Roller

The Bee-eater (*Merops apiaster*) and the Roller (*Coracias garrulus*) have in common the brilliance of their plumage, with its turquoise, yellow and orange. They leave southern and central Europe for their winter quarters in Africa. The first passes down the Nile and makes for the south-east of the continent, while the second spreads into the southern half of Africa.

The Hoopoe

The Hoopoe (*Upupa epops*) is one of the most striking of the European birds, with its pinkish breast, black and white barred wings and, when it condescends to spread it, its large fan-like crest (the bird is known as *Huppé* in French, i.e. crested, but it really derives its name from its cry of *houpou houpou*). From August it leaves us for the area to the south of the Sahara, but it returns again in mid-February. While the nesting area does not extend beyond France to the north, it stretches eastward right the way to Russia.

Woodpeckers

Among the *Picidae* or woodpecker family, the curious Wryneck (*Jynx torquilla*) is difficult to observe but is a migrant. Its nesting area covers a good part of Europe but its disappears southwards to places between the Mediterranean and the Equator.

Larks

A well-known bird in Europe is the Skylark (*Alauda arvensis*), a light-brownish bird that pours forth a delightful high-pitched song in both winter and summer. It is a partial migrant and in winter-time populations in France, for example, receive recruits from those to the north and east. The Woodlark (*Lullula arborea*), with its even more melodious song, presses on a little further south, while the Short-toed Lark (*Calandrella brachydactyla*), a rarer bird, reaches north-western Africa.

Swallows

We now come to the most famous migrant of them all. Curious fables have been woven round the swallows in the past, for the absence of the birds in winter-time required explanation.

From many parts of Europe one comes across a notion that swallows pass the winter at the bottom of ponds, sunk in a sort of torpor. Thus Regnard, in his *Voyage to Lapland*, wrote:

The Lapps make holes in the ice from place to place, and push by means of a pole that goes under the ice, their lines from hole to hole and likewise draw them out. But what is most surprising is that they find on their lines swallows that have clutched some small piece of wood in their claws. They appear dead when taken from the water and show no signs of life; but when placed near a fire they begin to feel the heat, recover a little, then spread their wings, and start to fly as they do in the summer. This peculiarity was confirmed by all that I asked.

This has never been verified scientifically, but one cannot fail to be struck by the number of similar stories from various parts of Europe. We can quote, without comment, one particular story. A swallow, it appears, fell into a vessel full of water but was brought back to life once more after being submerged for several hours. More recently it has been found that swallows deprived of food (the multitude of insects that make up the aerial 'plankton') through a sudden cold spell took refuge in shelters and entered into a state of suspended animation (internal temperature, respiration and heart-beats dropping). At present there are very few cases of real hibernation known in birds. One example is the North American Poor-Will (*Phalaenoptilus nuttallii*) which has been found in the foothills of the Sierra Nevada huddled in rock crevices with no detectable heart-beat or breathing and a body temperature of some 41°C below the normal (59–60°C). The Hopi Indians have long referred to this species as Hölchko, i.e. the sleeper. The Trilling Night Hawk (*Chordeiles acutipennis*) may also fall into this torpid state.

To return to swallows, there are ninety-five species in the world, of which five frequent Europe (the fifth, the Red-rumped Swallow, *Hirundo daurica*, is very rare).

The ordinary Swallow (*Hirundo rustica*) can be recognized by its long tail feathers and its reddish breast, the House Martin (*Delichon urbica*) by its notched tail, white rump-patch and completely white hinder parts. The Sand Martin (*Riparia riparia*) is brown, not black, above and has a brown band across the chest. The Crag Martin (*Ptyonoprogne rupestris*), which is much more rare and in France is found only in the extreme south, is brown above, has no breast band, and the tail is squared off and bears two white squares on it. All these species are migratory.

The autumn migration of the swallow (*Hirundo rustica*), unlike that of the swift, extends over quite a long period, from August to October. After travelling all day, the birds rest at night in 'dormitories', which are often in large reed-beds. Here one can count thousands of swallows – and see tens of thousands. It is to

Africa that they return, principally southern Africa, the Congo and West Africa. Depending on their country of origin, they have a more or less marked preference for a particular part of Africa for their winter quarters (see Fig. 14). Unlike the swift, which arrives each year at the same time with the regularity of a metronome, the swallow only migrates when conditions are suitable and the temperature sufficiently high. It is, in fact, temperature more than any other factor that determines the progress of the migratory front. Thus, if one plots the front during the course of the return journey to Europe one obtains a series of lines which von Middendorf named 'isopites' (later named isochronal lines). Henry Neville Southern placed on a map the isochronal lines for a number of migrants and found that they corresponded remarkably well with the advance of isothermal lines (i.e. lines connecting places with the same temperature). The swallow's progress northwards shows a close correlation with the 9°C isotherm, although the swallow tends to move slightly in advance of the isotherm. This explains why the western part of Europe, with its more equable climate, is recolonized before the eastern part, which is subject to the rigours of a continental climate. Swallows have also been found to follow this 9°C isotherm back southwards on their return journey. This is a very clear example of a migration that comes under the direct influence of external conditions.

The house martin also migrates to Africa after it has raised two or even three broods. It occupies almost the same wintering area as the swallow and also avoids the great equatorial forests. It returns to Europe a little later, in April or May.

The sand martin (*Riparia riparia*) forms large colonies during the nesting season and is found throughout Europe. It makes for eastern Africa (and also, in fewer numbers, for Lake Chad, Sierra Leone and southern Africa). Leaving much earlier in the autumn than the other species, it returns to Europe between March and May.

The crag martin (*Ptyonoprogne rupestris*) is partly sedentary, i.e. in those parts of southern Europe where the climate is mild, but it migrates from countries that become too cold in winter (Switzerland, Montenegro). Its absence in winter is fairly brief, for it winters in North Africa up to the Sudan, and from February the first migrants begin their return journey.

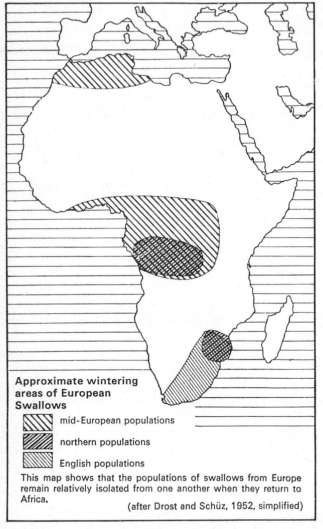

Approximate wintering areas of European Swallows

mid-European populations

northern populations

English populations

This map shows that the populations of swallows from Europe remain relatively isolated from one another when they return to Africa.

(after Drost and Schüz, 1952, simplified)

Fig. 14

Golden Oriole

The Golden Oriole (*Oriolus oriolus*) is a unique European representative of an otherwise tropical family with many species. It is certainly one of the most beautiful birds of our forests, as much for its superb song as for the brilliant yellow and black plumage of the male. From the end of July it sets off for the south-east, and it is one of the rare European migrants that takes that direction. It travels at night and seems to find no difficulty in crossing high mountain chains. Leaving early in the season, it is not pressed by adverse weather and, on arrival in Greece, it stops for a month to feed on figs; it then goes on to Egypt and Libya, and finally arrives in eastern Africa in October, where it winters. In spring it returns through North Africa, across the full width of the Mediterranean, and arrives in Europe in April and May after having traced a huge loop between the two continents.

Crow-like birds (Corvidae)

The crows of western Europe are semi-sedentary (i.e. in Belgium, Great Britain, France). This is not the case with other European populations, which carry out migrations with a more strictly east–west than north–south axis. In winter the number of Rooks (*Corvus frugilegus*) – recognized by the bare, whitish patch at the base of the beak – is considerably augmented by massive flights from the east (Russia, Baltic countries, central Europe). They are a great nuisance to farmers because they scratch up the seeds that they have planted, whereas in summer they are extremely useful in keeping down grubs and insects. The great flocks that come into France from the east do not mix with the local populations but repair to their own nesting colonies towards April. The Carrion Crow (*Corvus corone corone*) and the Jackdaw (*Corvus monedula*) are also partial migrants and, as is often the case, it is the young who are the most active migrants.

The Nutcracker (*Nucifraga caryocatactes*) can be recognized by its dark plumage and white rump. If one gets close enough to a nutcracker at the top of a pine tree one sees a crow-like bird a little larger than a starling with a long and specialized beak like a nutcracker and a plumage of a dark chocolate colour boldly speckled with white. It is not a regular migrant but from time to time it makes what can only be termed 'invasions'. That is to say, suddenly, one year, a mass of birds appear in countries where they are not normally found, only to disappear after some months, victims of unaccustomed conditions and the rigours of the journey. The cause of these invasions has been found to be connected with feeding. The nutcrackers leave Siberia (those of the Alps are sedentary) and they make their way to western Europe. The life of the nutcracker is very much tied to the fruiting of the arrolla or parasol pine (*Pinus cembra sibirica*), which forms the principal food of this bird. However, the fruiting of this tree can be very irregular, with good years followed by bad ones. Likewise, during the good years the nutcrackers flourish and raise a full brood of young, but during the lean years this greatly enlarged population of nutcrackers outstrips the food supply and suffers accordingly. It is at this time that the invasions occur, and they often prove to be a mass exodus without return, ending in the disappearance of a large part of the population.

The Jay (*Garrulus glandarius*), a sedentary bird, sometimes collects in great numbers and takes part in movements which are as yet little understood.

Titmice

These birds, though small, are remarkably aggressive. Place two birds in a cage, one of them a titmouse, and the latter will soon pull out the feathers of its unfortunate companion and kill it. Those who ring birds are well aware that the Blue Tit (*Parus caeruleus*), a determined little predator, will attack and in the most cruel fashion kill a bird five times its own size.

In Britain the majority of titmice are sedentary, but those that live in the east will sometimes go as far as Russia, as has been found from ringing experiments with the Great Tit (*Parus major*) and the blue tit. In Switzerland it has been estimated that 15 per cent of the juveniles and 10 per cent of the adults of great tits exhibit a migratory behaviour. The description 'partial migrant' is thus a very apt one for these species. The Coal Tit (*Parus ater*) regularly makes an autumn flight through the valleys of the Jura Mountains and the Alps, and the populations of eastern Europe show a widely erratic behaviour.

The Wall Creeper (*Tichodroma muraria*), which belongs to the family *Sittidae*, is a lovely little bird with a long, thin beak and pearly grey plumage enriched with two areas of carmine on the wings. This bird frequents rock faces and if you happen to see it in a town it will be on the walls of some large building such as a church. It is rare and is found in mountainous regions, ascending up to 12,000 feet during the summer. This is a particularly interesting bird because it affords an example of altitudinal migration. When winter comes it does not head southwards but descends from the heights down into the valleys and may spread on to the plains without any preference either for the north or the south.

Turdidae (Thrushes, Blackbirds, etc.)

This is an important group of species, some of which are among the commonest and best known in Europe. These are mostly partial migrants. The four most common species of thrush are no exception and in the winter birds from the north and from the east (Russia) augment the existing populations in western Europe.

The Redwing (*Turdus musicus*), which has a prominent white 'eyebrow' and red under the wing, migrates at night and at this time one can hear their thin cries even over towns. They have a migration route which is aligned north-east by south-west, as also do the Mistle Thrush (*Turdus viscivorus*), with a light breast strongly marked with black spots, the Song Thrush (*Turdus philomelos*), with smaller spots and buff wing-linings, and the Fieldfare (*Turdus pilaris*), with a grey head and rump and a rust-coloured back. An erratic behaviour in the winter quarters (western Europe, Spain, Italy, North Africa) is superimposed on the seasonal migrations. The fieldfare and the mistle thrush then move in cautious flocks searching out the berries which are their favourite food. The mistle thrush plays an important role in the dissemination of mistletoe, for it is extremely partial to the berries.

The Blackbird (*Turdus merula*) is also a partial migrant; sedentary in western Europe, it becomes more and more migratory as one proceeds north and eastwards. In general, the juveniles and the females are more migratory than the males, a phenomenon which is, however, not unusual.

The Ring Ouzel (*Turdus torquatus*), which has a striking white stripe across its throat, is common in the French Alps. In winter it descends and flies as far as North Africa. The northern sub-species (Scotland, Switzerland) oscillates between its nesting area and its winter quarters (Spain, North Africa). The Rock Thrush (*Monticola saxatilis*), a fine-looking bird with a blue head and orange breast in the male, reaches considerable heights (over 8,000 feet in the Alps) and winters in tropical Africa.

The Wheatear (*Oenanthe oenanthe*) can be recognized in flight by its grey back, beige front, black wings and distinctive white rump. Approaching it on the ground, it will throw you a nervous call, a hard *chack, chack-weet* or *weet-chack*. This bird is an extra-ordinary migrant. There are different subspecies of wheatear nesting in Europe, Greenland, the whole of Russia and the north of Canada. It may seem surprising that Canada is mentioned in a section dealing with European birds, but this is the extraordinary thing about the wheatear. All the birds, whether from Europe, Siberia, or the north of America and Canada, spend the winter in tropical, central or southern Africa. If the journey is long enough for the birds from western Europe, it is quite incredible for those Siberian and North American populations, for the latter must cross the Bering Straits, traverse the whole of the Soviet Union and then pass southwards to Africa. This represents a round trip of some 20,000 miles each year. During this journey the density of wheatears grows enormously and along the Atlantic coasts men trap them in enormous quantities (on some days tens of thousands of birds are caught).

The Stonechat (*Saxicola torquata*), in which the male can be identified by its black head, orange-red throat and white patches on wings and neck, is a partial migrant wintering in Spain and North Africa.

In its breeding plumage the Redstart (*Phoenicurus phoenicurus*) is a beautiful little bird with a pearly grey back, white eyebrow, black face, an orange breast fading to white and a red tail, all harmoniously blended. Its migration begins in August and reaches a peak in mid-September. Nesting throughout Europe, the redstart travels regularly to the south-west in autumn. It continues right up to Africa and then spreads from the Mediter-ranean coast of Africa down to the Equator. (Some, however, are known to winter in Europe). The Black Redstart (*Phoenicurus ochurus*) is much darker, with sooty black underparts relieved by

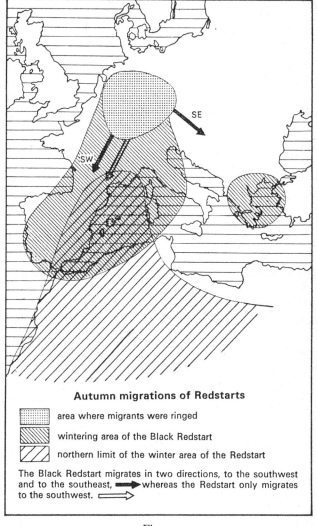

Autumn migrations of Redstarts

:::: area where migrants were ringed

▧ wintering area of the Black Redstart

▨ northern limit of the winter area of the Redstart

The Black Redstart migrates in two directions, to the southwest and to the southeast, ➡ whereas the Redstart only migrates to the southwest. ⇨

Fig. 15

the constantly flickering rusty rump and tail, earning it the name of '*ramoneur*' in French (chimney-sweep). This bird is a partial migrant and wintering populations can be found along the south coast of England as well as in France, Spain, the Balkans and Greece. Among the migrants there are two schools; most migrate to the south-west (as do the majority of passerines), but others travel south-east, towards Greece, Turkey and Egypt.

The Nightingale (*Luscinia megarhynchos*), the most famous of all European song birds, is shy and secretive and avoids intruders. The chestnut-coloured tail merges into the buff brown of the back, while the breast is plain. Its nocturnal habits are shown by its large round eyes. The period of singing ends in mid-June and only through ringing can we be certain that it is still with us until it migrates in September. It crosses the Sahara and makes for the tropics (see Fig. 16). The Thrush Nightingale (*Luscinia luscinia*), which nests in eastern Europe, winters a little further south than the nightingale.

Another musician is the beautiful Bluethroat (*Cyanosilvia svecica*), which in the male has a bright blue throat-patch with black and chestnut bands below and conspicuous chestnut panels at the base of the tail; a white patch in the middle of the blue of the throat adds to the regal effect. It has a melodious song, a little like that of the nightingale, but most of the time it is a secretive bird that flies rapidly between beds of rushes and one only sees the brown of the back and the characteristic chestnut tail. A summer visitor (although there are indications that it may winter in western France), there are several distinct subspecies that nest in Europe. The Scandinavian subspecies has the white of its throat-patch replaced by chestnut. After the local populations of western France have departed for North Africa, the northern birds begin their migration in September.

The Robin (*Erithacus rubecula*) is also an example of a partial migrant. The Scandinavian populations are entirely migratory, but as one goes south-west one finds that sedentary birds begin to predominate, especially males. In winter, France receives an important contingent of migrants coming south from countries between Belgium and the Soviet Union. Some of them continue their flight to the south-west and cross the Mediterranean, while others from Germany make for Egypt and the Sudan.

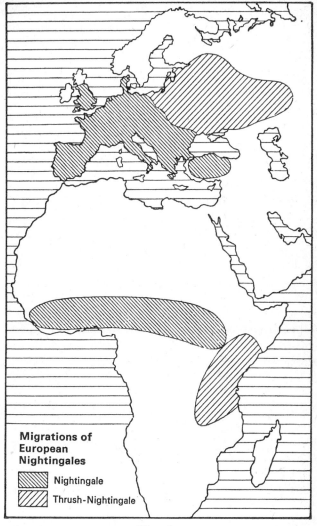

Migrations of European Nightingales

Nightingale

Thrush-Nightingale

Fig. 16

Warblers, Flycatchers, etc.

This is another group that is rich in species in Europe. In contrast to the robust roundness of the thrush family, the Warblers (family *Sylviidae*) are slender and delicate and have a thin beak and softer and more diffuse colours (reddish-browns, grey-greens, pale yellows). They are almost all long-distance migrants, in spite of their small size, and although they make up a large part of our fauna in the summer, only a few stay with us during the winter. Those that remain, moreover, and leaving aside the Blackcap (*Sylvia atricapilla*), are timid birds that are difficult to observe, like the Dartford Warbler (*Sylvia undata*) found in Brittany, or the Cetti's Warbler (*Cettia cetti*).

Of the warblers that reach southern Africa, one can cite the Great Reed Warbler (*Acrocephalus arundinaceus*); the Marsh Warbler (*A. palustris*) – very difficult to distinguish from its companion the Reed Warbler (*A. scirpaceus*) which leaves Europe later and does not reach so far south; also the pretty Sedge Warbler (*A. schoenobaenus*), with its odd explosive cry that becomes a stuttering rattle when excited, the Barred Warbler (*Sylvia nisoria*) and the Garden Warbler (*Sylvia borin*) so common in our thickets.

Tropical Africa draws another set of warblers. There is Savi's Warbler (*Locustella luscinioides*), which looks rather like a large reed warbler but has a song like the grasshopper warbler, and also there are the Icterine Warbler (*Hippolais icterina*) and the Polyglot Warbler (*H. polyglotta*) which do not overlap in Europe, the first nesting in Spain and western France, and the second nesting in the remaining areas. Mention should be made also of the Whitethroat (*Sylvia communis*), a pretty warbler with a distinct white throat, and the Subalpine Warbler (*S. cantillans*) which also winters in the savanna regions of the southern Sahara.

Those who have to ring warblers soon get to know their different species, but for the beginner it is something of a nightmare, for the differences in plumage are often very slight. Identification by their songs is sometimes essential. Those who have failed to recognize the Chiffchaff (*Phylloscopus collybita*) can probably pick out its characteristic call of two notes irregularly repeated, sounding more or less like its name. This little bird, weighing only six grammes, spends the winter from southern

Europe to the southern Sahara. The Willow Warbler (*Ph. trochilus*), which is scarcely any larger, travels as far as the Cape of Good Hope after crossing the Sahara in sixty hours of flight.

Leaving the warblers, we pass on to the passerines. The Firecrest (*Regulus ignicapillus*) is also a tiny bird (about five grammes in weight) and is a partial migrant. One curious aspect of its behaviour is that some of the birds winter in the south of England even though this is to the north of their nesting area.

The Hedge Sparrow or Dunnock (*Prunella modularis*), a common bird of gardens and hedgerows, is not a sparrow at all; it should not be confused with the Tree Sparrow (*Passer montanus*) or the House Sparrow (*Passer domesticus*), both of which have a prominent black 'bib' and belong to the family *Passeridae* not the *Prunellidae*. The hedge sparrow is a partial migrant which sometimes travels hundreds of miles south-westwards when driven by cold weather.

The Spotted Flycatcher (*Muscicapa striata* – family *Muscicapidae*) is an elegant little bird. Having caught an insect it will repair to its perch with the creature between its beak and, once it has settled itself, the grey upper parts and the finely striated lower parts will confirm its identity. In winter it heads for tropical and southern Africa. To do this but to avoid crossing the Mediterranean, the population splits into two parts, one migrating south-west over Gibraltar, the other passing down from Scandinavia and central Europe and across Italy and Greece. The Pied Flycatcher (*Ficedula hypoleuca*), with a more contrasted black and white plumage, migrates to the south-west and reaches central Africa.

The Tree Pipit (*Anthus trivialis* – family *Motacillidae*) lives in the large trees and fields that border woods. It is a bird with a slender beak, olive above and lightly striped below. It leaves Europe in September, crossing the Sahara and making for central Africa.

Wagtails are always a cheerful sight. One of them, the Blue-headed Wagtail (*Motacilla flava*) is a very high-flying migrant. There are several subspecies in Europe but all have a superb yellow breast and belly. They migrate in flocks often of a considerable size (in autumn some tens of thousands are often seen daily along the shores of the Caspian). Mountain peaks of

9,000 feet do not stop them in their journey to Africa, which they colonize right down to the extreme south.

The Waxwing (*Bombycilla garrulus* – family *Bombycillidae*) is a curious bird. Huge invasions of waxwings take place periodically, the last one being in 1966. This is a plump little bird with a long pinkish crest and yellow tip to the tail. The invasions, which take place about every ten years, are superimposed on a regular migratory pattern. The cause of the invasions, which bring the waxwings to areas where they are rarely seen otherwise, remains obscure.

The male of the Red-backed Shrike (*Lanius collurio* – family *Laniidae*) is about the size of a sparrow with a curved beak, a grey head with a black band across the eyes and a chestnut back. This small predator exhibits what we should consider a cruel nature, for it makes a kind of larder by impaling insects on the thorns and spikes of trees, to be consumed later. This bird is a strong migrant and Fig. 17 shows how the wave of migrants from Europe concentrates over Greece in its passage to southern Africa. The Eurasian populations take a different route and can winter in areas much closer to their nesting area and still enjoy a tropical climate (in the Indies, for example).

Starlings

These are typical partial migrants. Those of the British Isles and western Europe are erratic in their behaviour, while those from the north and the east are more strictly migratory. Flocks of Starlings (*Sturnus vulgaris*) are quite independent of one another and rarely mix. The different populations have different winter ranges according to their place of origin. As regards the migration of juvenile starlings (see page 45), this is not a true migration but a radial dispersal from the nesting sites.

Finches

Most of the finches are common and well-known species; they are birds without a strongly developed migratory behaviour. Those of them that do leave the country do so for the south-west to enjoy the warmer climates of Italy, the south of France and the Iberian peninsula, such as the greenfinch, chaffinch etc.

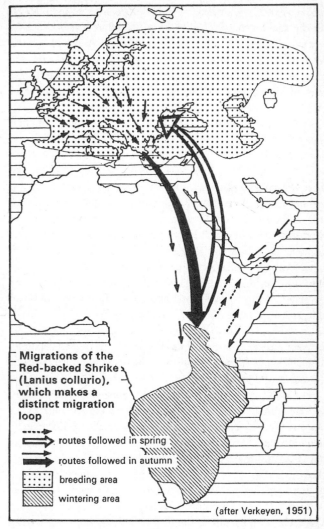

Migrations of the
Red-backed Shrike
(Lanius collurio),
which makes a
distinct migration
loop

→ routes followed in spring
➡ routes followed in autumn
⋮⋮⋮ breeding area
▨ wintering area

(after Verkeyen, 1951)

Fig. 17

Others make for North Africa, as for example the Linnet (*Carduelis cannabina*), the Serin (*Serinus canaria*), the Siskin (*Carduelis spinus*) and the Hawfinch (*Coccothraustes coccothraustes*). They are not dismayed by mountain ranges, and one can see the Goldfinch (*Carduelis carduelis*), redpolls and linnets at heights of 9,000 feet. These birds are often inconsistent and the flock of linnets which enliven a Breton landscape with the cries of carmine-breasted males may perhaps be found next year in the same place but the following month might be in Seville.

The finches can be dealt with more fully. The Brambling (*Fringilla montifringilla*) can be recognized in the distance by its white rump and rusty chest and shoulders, contrasting with its sooty head and wings. It nests in birch and conifer forests of the subarctic countries. In October the first flocks arrive in our region (the periphery of the Mediterranean being colonized after that) and they then exhibit an erratic behaviour. When they find abundant food they will congregate in immense flocks. In the evening, the whole flock congregate like starlings in a nearby copse and such 'dormitories' present an astonishing spectacle. In one instance in Switzerland, some 30,000,000 brambling were estimated to have congregated every evening. In the winter of 1964–5 I myself saw a flock of 1,000,000 finches spread out in the salt marshes of Anse d'Aiguillon in the Vendée. Such phenomena are very irregular, and in a locality where every step sets up a flight of finches one may find that the density of birds is quite normal the following year.

The Latin name for the Chaffinch is *Fringilla coelebs*, the word *coelebs* meaning bachelor. This stems from the fact that in the chaffinch, which is part migratory and part sedentary, the males predominate. The chaffinches of north-east Europe are more migratory in behaviour and they reach as far as North Africa. If the wind is favourable the females readily cross the Channel and descend on England, whereas the males apparently hesitate to fly over water and accumulate along the Dutch coast.

A rather curious bird is the Crossbill (*Loxia curvirostra*), both in its appearance and in its habits. The male (dull red with dark wings) and the female (yellowish grey) both have beaks like a parakeet, with the tips of the mandibles crossed. This type of beak is specialized for feeding on pine cones. The crossbill is unique in Europe since it is not necessarily obliged to breed in spring but only when it finds the right feeding conditions, the birds wandering from place to place until they achieve this. In some years this wandering takes on the proportions of an invasion, the crossbills spreading across the countryside in their desperate search for food. Like other such invasions (snowy owls etc.) there is no return journey and the result is the establishment of a new population level in keeping with the food supply.

Buntings

Most of the buntings of our region are sedentary or else erratic wanderers. Only the Ortolan Bunting (*Emberiza hortulana*),which nowadays has a rather restricted range, is a true migrant, making its winter quarters in Ethiopia and Somalia. France occasionally receives two of the northern nesting species, the Snow Bunting (*Plectrophenax nivalis*), which appears from time to time on the northern coasts of France, and the Lapland Bunting (*Calcarius lapponicus*). The first of these nests some years later in the Faroes.

2 North America

Unlike their compatriots of the Old World, the North American migrants do not have to overcome such large obstacles as the Mediterranean or the Sahara during their flights south. The Gulf of Mexico is certainly a barrier but it is not unavoidable, for birds can circumvent it either by flying across Mexico or by using the West Indian islands as stepping-stones. There are, however, certain species that spurn this easy solution and head straight across the Gulf. In North America, again unlike Europe, the mountain chains and the river drainage patterns are more gradual and this enables large numbers of tropical birds to make their way north. North America has thus been colonized in the past by birds with tropical affinities such as the Tanagers (family *Thraupidae*) and the Hummingbirds (family *Trochilidae*).

Under such conditions, the migration routes are relatively simpler than those of Europe, and Frederick Lincoln has mapped out six major routes, which he termed 'flyways'.

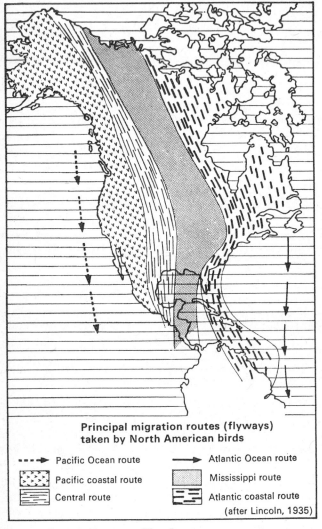

Principal migration routes (flyways)
taken by North American birds

- - - ▶ Pacific Ocean route   ⟶ Atlantic Ocean route

Pacific coastal route   Mississippi route

Central route   Atlantic coastal route

(after Lincoln, 1935)

Fig. 18

1 Atlantic Ocean Route

This flyway is over the open ocean and is used only by birds with unusually well-developed powers of flight, notably shore birds. The best example is the Golden Plover (*Pluvialis dominica dominica*), a bird which certainly upholds the migratory tradition of the plover family, making a journey of not less than 12,000 miles from one year to the next (Fig. 19). Although it has not been proved yet, it is possible that these birds sometimes rest on the water. The adult golden plovers are a good example of birds that make a 'loop' migration. Leaving the coast of Labrador they fly over the ocean (passing outside the line of the West Indies) and do not make contact with the land again until they have reached the level of Brazil, having crossed some 2,500 miles of open sea. Continuing southwards, they finally stop only when they have reached the southern part of Brazil or Uruguay. In spring the plovers return, but this time they take the overland route on their way back to the northern tundra regions, thus describing an enormous loop. The young, however, take the overland route both ways.

2 Atlantic Coast Route

This flyway, which begins in the far north-east of the continent, gathers tributaries from the north and north-west as it passes southwards. Many of the web-footed birds take this route, chiefly because of the abundant rivers and stretches of water along the way. The birds can continue southwards over Cuba and the West Indies and, by means of a 600-mile flight over the sea, reach the northern coasts of South America, although few do this. Among the adventurous, however, the best known is the Bobolink (*Dolichonyx oryzivorus* – family *Icteridae*), a bird of tropical origins which nevertheless nests in Canada and the north of the United States. When winter comes the bobolink heads southwards along the Atlantic Coast Route, crosses the Caribbean area and winters in Bolivia, Brazil and Argentina, returning to its northern nesting sites the following spring. Lincoln has termed this the Bobolink Route.

One of the warblers, the Blackpoll (*Dendroica striata*), has a long migratory route that passes through the West Indies. On its

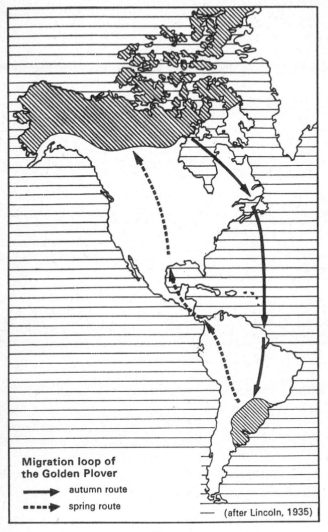

Migration loop of the Golden Plover

→ autumn route

▪▪▪▶ spring route

— (after Lincoln, 1935)

Fig. 19

return from South America (Guyana, Brazil, Peru, even Chile), the blackpoll leaves Venezuela and passing over Cuba arrives in the United States, where it heads north and then north-west, even as far as Alaska, a total journey of some 5,000 miles.

3 Mississippi Route

From Alaska to the Gulf of Mexico, this flyway collects a great number of birds which are drawn into the flyway like the tributaries of a river. Dotted with lakes, marshes and rivers, it attracts a large number of water-fowl, especially mallard, pintails and Canada geese. It brings to the Gulf of Mexico a good many passerines, including the well-known American Robin (*Turdus migratorius*), which is an accidental visitor to Europe, and American flycatchers (family *Tyrannidae*) like the Black Phoebe (*Sayornis nigricans*), which is sedentary in the north but nomadic in the south. About sixty species actually cross the Gulf of Mexico and reach Yucatan. The case of the Ruby-throated Hummingbird (*Archilochus colubris*) is quite extraordinary. This tropical bird goes as far as Labrador to breed and on its return southwards it crosses the Gulf in a single flight, which is amazing when one considers the tiny size and apparent fragility of the bird.

Another bird with tropical affinities is the Scarlet Tanager (*Piranga olivacea*) which nests in the eastern United States and Canada. The population concentrates into a very narrow front as it moves southwards and finally crosses the Gulf of Mexico to Panama and then spreads out once more in Colombia, Ecuador and Peru. Although certain swallows do the same (the Barn Swallow, *Hirundo rustica erythrogaster*, for example), others such as the Cliff Swallow (*Petrochelidon pyrrhonota*) hug the land and in so doing make a detour of some 2,000 miles.

During the spring migration, the salt marshes of Louisiana collect vast numbers of geese and ducks. One author, writing of the blue geese, gave the following description:

We found that the flock was compact and filled an area of two and a half miles by 500 to 700 yards wide. While we were going through this on horse-back, the birds were so tame that they did not fly up but merely parted on either side of our path and during the whole time we went through the flock there were geese on either side of us only three or four yards away. We estimated that the flock comprised 1,250,000–1,500,000 geese.

4 Central Route

This flyway includes the crossing of the Rocky Mountains, but it avoids the Gulf by passing down through Mexico. Certain populations of barn swallow follow this route and reach South America by a purely overland migration. Another example of the astonishing hummingbirds is provided by the Rufous Hummingbird (*Selasphorus rufus*) which nests as far north as Alaska, crosses over the Rockies at heights of 9–12,000 feet and then winters in the high Mexican plateaux. Like the Atlantic Coast and Mississippi routes, one finds numbers of ducks using this flyway.

5 Pacific Coast Route

This is a less well-defined route than the preceding ones. It seems to have a number of main stems and numerous tributaries. Here in the western United States the climate is milder and the birds show less tendency to migrate. Here again, one notes the abundance of ducks and geese and these winter on the lakes of southern California.

6 Pacific Ocean Route

This flyway is analogous to the first (Atlantic Ocean Route), and it is also used mainly by shore birds. As on the east coast, there is a plover, the Pacific Golden Plover (*Pluvialis dominica fulva*) that is characteristic of this flyway.

The Pacific golden plover nests in northern Canada, Alaska and Siberia. In autumn it sets off on a flight that takes it to the Hawaiian islands after some 2,065 miles over the sea. Not content with this feat, it continues on to some of the islands of Oceania. In following the Asiatic coast and pressing on southwards it reaches Australia and New Zealand.

The peregrinations of the Bar-tailed Godwit (*Limosa lapponica baueri*) are quite astonishing. Having nested in Alaska it heads for New Zealand and Australia and returns to Alaska in the following season.

The Bristle-thighed Curlew (*Numenius tahitiensis*), first discovered two centuries ago by Captain Cook on his first visit to

Tahiti, is not a wandering bird in the strict sense but spends a large part of its life over the seas. Breeding in Alaska, these birds make a huge arc across the Pacific to Polynesia to spend the winter.

It should be understood that these six flyways are the routes that birds take for preference but they are not train-lines, and it is not unusual to find a bird changing over from one flyway to another.

Turning briefly to the winter distribution of the North American birds, it is noticeable that central America receives the vast majority of them. The density of migrants is here higher than in any other part of the world (their density is always greater than that of the local birds, and the result of this is that some of the latter wait until the migrants have departed before they themselves breed). Unlike Africa, where a number of birds reach the very tip, the extreme south of South America harbours rather few migrants from the north. In Chile, for example, only five purely terrestrial and twenty-nine shore birds have been recorded wintering. A large number of birds from North America do not, in fact, cross or skirt the Gulf of Mexico but concentrate in the countries bordering the Gulf.

3 Tropics and Equatorial Regions

In tropical regions it is not so much variations in temperature (which are in any case slight) but variations in rainfall and humidity that determine the migrations. For birds which are not dependent on the flying insects that comprise the aerial plankton, the absence of dire climatic factors such as rigorous winter gives way to other and often curious motives for migration. The Wood Stork (*Mycteria americana*), for example, journeys between the Orinoco and the Amazon basins in accordance with changes in water levels between the two, always making for the area where the water level is lowest, and thus provides the best feeding opportunities. The equatorial forests, where temperature and humidity are almost constant, together with the highly specialized type of habitat, have led to mainly sedentary birds.

Central America, which is invaded by very many species of birds from North America, has relatively few true migrants of its own. One can cite, however, the Yellow-green Vireo (*Vireo flavoviridis*) which nests in Mexico and winters in the upper Amazon basin. Migratory behaviour is here of a rather erratic nature and where the factors that trigger it are cyclic (food abundance, for example), this erratic behaviour can have all the appearance of a true migration.

Turning to tropical Africa, one finds that the alternation of wet and dry seasons is the main factor provoking migration. But, unlike migrations in the northern hemisphere, these in the tropics are not strictly in one direction only. One cannot speak of 'wintering' and 'summering', and the directions of migration must be related to the place where nesting occurs. In the northern hemisphere the migrations are all of one kind since the birds are driven southwards to warmer areas by the approach of winter. In Africa, on the other hand, one finds migrations going in opposite directions or crossing each other, some birds leaving their nesting area while others are returning to theirs. The movements of two such species are thus identical but, as defined by their reproductive cycle, are exactly opposite. Dr James Chapin has given an excellent example of this in two species of nightjar of Africa. The Standard-wing Nightjar (*Macrodipteryx longipennis*) is found along a broad belt of equatorial forests from Kenya to Senegal, where it breeds in the dry season and moves northwards during the rains. The Plain Nightjar (*Caprimulgus inornatus*) lives in a similar belt but further eastwards and it breeds in the northern and drier part of its range only during the wet season, moving southwards during the dry season.

Another nightjar, the Pennant-winged Nightjar (*Cosmetornis vexillaris*), which resembles the European nightjar but is equipped with long flight feathers which prolong its wings some eighteen inches behind it, makes transequatorial migrations. After nesting in the region to the south of the great equatorial forest, it then moves northwards following the rain belt and thus following the emerging insects on which it feeds. The Abdim Stork (*Sphenorhynchus abdimi*), on the other hand, nests in an area to the north of the Equator and then goes southwards at the close of the rainy season.

In the Far East the climatic conditions differ widely from country to country and the multiplicity of factors involved does not result in very regular or clear migration patterns. India receives wintering birds from Eurasia which have crossed the Roof of the World, and it is strange that ducks, geese and shore birds are able to adjust themselves to altitudes of up to 22,000 feet. There are about forty birds that nest in Siberia and spend the winter in India.

Speaking of the Himalayas, one of the best examples of altitudinal migration is the White-capped Water Redstart (*Chaimarrornis leucocephalus*). This bird nests in the eastern Himalayas at heights of as much as 16,000 feet, but when winter comes it descends to the valleys at about 1,800–7,500 feet.

4 Southern Hemisphere

As far as bird migrations are concerned, the two hemispheres cannot be considered equivalent.

While numbers of northern species winter in temperate regions in the southern hemisphere, the reverse is not true, and while some rare southern species cross the Equator, they do not reach further north than the Tropic of Cancer. One must, however, except certain sea birds, such as the Greater Shearwater (*Puffinus gravis*) which nests in Tristan da Cunha and then returns to our area.

The indigenous bird fauna, which consists of only a small number of species, is for the most part sedentary, for the winters are not very rigorous in the southern temperate regions (it should be remembered, too, that the temperate regions of the southern hemisphere are not nearly so extensive as they are in the north). In Africa, it is rainfall and humidity that are the overriding factors as far as migration is concerned, just as they are in the intertropical regions. There are exchanges between Madagascar and the mainland, and the Madagascar Squacco Heron (*Ardeola idae*), which nests on the island between October and December, returns to Africa from June until October. The Madagascar Cuckoo (*Cuculus poliocephalus rochii*), the Broad-billed Roller (*Eurystomus glaucurus glaucurus*) and certain others do the same, but on the whole the birds of Madagascar are sedentary. As mentioned earlier, many European species, such as shore birds, terns and swallows, reach as far as Madagascar in winter.

The situation is a little different in South America and Australia where contrasts in climate are rather more marked. The extreme south of South America has quite hard winters and some of the birds there are long-distance migrants. Some travel far to the north, as for example the Southern Brown-chested Martin (*Phaeoprogne tapera fusca*) which, nesting in Tierra del Fuego, winters in the Amazon, Venezuela and even as far as Mexico. However, they are exceptional, and only a small minority of birds pass north of the Equator. The well-known Ashy-headed Goose (*Chloephaga poliocephala*) moves northwards when the cold Patagonian winds blow, reaching central Argentina and central Chile, and their massive flights are sometimes very impressive.

Ranging from humid forest to barren desert, Australia has great climatic diversity and as a result one finds a number of migrants among the indigenous birds. For example, there is the Sacred Kingfisher (*Halcyon sancta*) which leaves the southern part of Australia in March and flies northwards to Malaya, New Guinea and the Solomon Islands. As elsewhere, the swallows and cuckoos are migratory, but the major part of the Australian bird fauna is more nomadic than truly migrant. Apart from the local birds, Australia receives an important contingent of northern travellers (above all shore birds) that must cover enormous distances to get there. One can cite the Spine-tailed Swift (*Hirundapus caudacutus*), which makes the long journey from China and Japan. From the south come a number of birds which can be termed pelagic (i.e. wandering) which take refuge in Australian waters from the Antarctic winter.

To the east lies New Zealand, a country dominated by its isolation which, in terms of migratory behaviour, results in a high degree of immobility. The generally mild climate also contributes to the insularity of the species found there. An exception is the Long-tailed Cuckoo (*Urodynamis taitensis*), which nests in New Zealand and then invades Oceania during the winter.

The Bronze Cuckoo (*Chalcites lucidus lucidus*), following the example of many sea birds, uses the prevailing south-east winds to help it on its way to the Solomon Islands. Certain shore birds and some gulls make a kind of aerial bridge between Australia and New Zealand. It can be noted also that New Zealand is used as a stop-over for such long-range migrants as the Bar-tailed

Godwit (*Limosa lapponica*) and the Knot (*Calidris canutus*), which come from the northern hemisphere.

5 Migrations of Sea Birds

Two principal groups of sea birds can be distinguished. First, there are those that keep near to the coasts and, if they move away from the coast, travel inland. The Black-headed Gull (*Larus ridibundus*) is an example. The second group comprises the birds of the high seas – the pelagic birds – which for most of the year are not tied to any one country or place.

The first group contains many members of the *Laridae*, a family that includes the gulls and terns. Of the gulls seen on our coasts, the most migratory is certainly the Lesser Black-backed Gull (*Larus fuscus*), a bird similar in size to the common Herring Gull (*Larus argentatus*) but with the legs yellow or orange (not flesh coloured) and the wings dark grey (tending to blackish in the Scandinavian subspecies). It is a partial migrant which winters from the Baltic but reaches as far south as Africa (Nile, Niger, Lake Victoria and the Congo). The more western subspecies of the Lesser Black-backed Gull (*L. fuscus graellsi*) nests in England and the Low Countries and on some islands off Brittany. A few of the British nesting birds actually winter in the British Isles but most follow the Atlantic coast southwards as far as Portugal, some even reaching West Africa. Sometimes, during these migrations, the gulls will fly far inland to regions apparently most inhospitable to sea birds. Thus the lesser black-backed gull has often been recorded from the Sahara. The Greater Black-backed Gull (*Larus marinus*) resembles the previous species in general colour pattern but has flesh-coloured legs and the wings and back are jet black; it is often seen perched on a buoy near a group of herring gulls. It frequently attacks the chicks of other species and will even go for young rabbits. Nesting from Scandinavia to Brittany, this species spreads down the whole Atlantic coastline, occasionally right up to the Mediterranean. The herring gull, whose numbers are increasing with the growing abundance of fish remains and other waste matter from the fishing industry – often at the expense of other species such as terns – is more of a wanderer than a true migrant, covering distances of up to 300 miles. The Common Gull (*Larus canus*) is similar to but smaller than the herring gull and has a greenish yellow beak and legs. It breeds in Scandinavia, along the Baltic coasts and in the north of the British Isles, and is found off the coast of France from August to April. To the south, it travels as far as Morocco, but in southern Europe it often flies some distance inland and has been recorded in Switzerland, having travelled up the Rhine.

The black-headed gull (*Larus ridibundus*) is distinctive with its dark brown head and crimson beak and legs; in winter the head is white with blackish marks before and behind the eye, but the characteristic long wedge of white on the primaries remains. This gull frequents both littoral and inland areas, where it nests in large numbers on lakes throughout Europe. After nesting it carries out a complex migratory pattern with a mixing of populations, except in England and the Mediterranean where the populations are sedentary. During its winter wanderings the birds often travel considerable distances, and there are cases of birds ringed in Russia being recaptured in Switzerland.

In summer-time, the inexperienced can easily confuse the Mediterranean Gull (*Larus melanocephalus*) with the black-headed gull, but the former has uniformly grey wings without a trace of black on the tips of the primaries. To see this species in breeding dress one must go to the shores of the Black Sea (where one can also see the Little Gull (*Larus minutus*) whose movements are shown on the map on page 97) or to Greece or Asia Minor.

Among the species which do not stray from the coasts are the Razorbill (*Alca torda*), the Guillemot (*Uria aalge*) and the Puffin (*Fratercula arctica*).

The razorbill nests in an area extending from the Arctic to sites off the coast of Brittany and in winter it migrates southwards, some birds rounding the Iberian peninsula and entering the Mediterranean. The guillemot looks something like a razorbill but has a noticeably longer but more slender beak. Its behaviour is also similar to that of the razorbill, but it has been found that the migratory impulse seems to dwindle with age. The curious puffin, with its great red, blue and yellow triangular beak, looks both comic and splendid. It breeds in colonies, from Scandinavia to the coasts of Brittany, and winters from the Channel to the Azores.

The Gannet (*Morus bassanus*) is a magnificent sight when it is

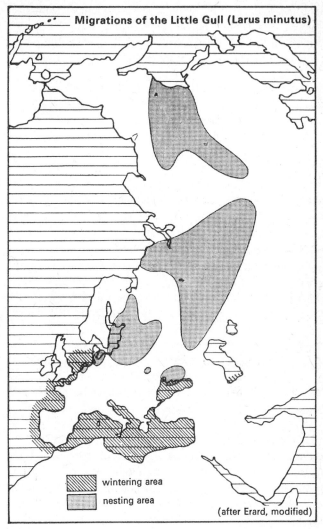

Fig. 20

diving for fish. The immaculate white plumage is heightened by the jet black of the outer parts of the wings. Suddenly, with wings half folded, the birds will drop out of a blue Atlantic sky, plunging sixty feet to break the surface with a splash and emerge with a fish. Rarely solitary, gannets nest in colonies and in the winter disperse along the coasts of Europe and as far as Senegal. The brown plumage of the immature birds only changes to the white of the adults when they are two years old; they are more adventurous than the adults and it is the latter that usually remain along European coasts.

Pelagic birds

These lords of the high seas, among which are some of the most famous migrants, cannot fail to excite our admiration. Most of them, unlike the birds that glide over the land, have long, thin, pointed wings and they use them to exploit differences in wind speed between air currents above and at the level of the waves. With the terns we come back to the family of *Laridae* mentioned earlier. The terns have a graceful and elegant flight unlike most gulls, from which they are further distinguished by their long and pointed beak and black 'skull-cap'. The Arctic Tern (*Sterna paradisaea*), which is difficult to distinguish from the Common Tern (*Sterna hirundo*), the commonest of our terns, is surely one of the most amazing of all migrants – a sentiment expressed elsewhere in this book, but it is difficult not to marvel at all these long-distance migrants. The Arctic tern nests in the north of the three northern continents (Europe, Siberia and North America), often in very high latitudes, although there are also some nesting sites further south (for example the island of Dumet in France). Leaving one pole for the other, a journey that requires flying twenty-four hours a day for eight months of the year, they spend the winter along the shores of the Antarctic, often well to the east and in the longitudes of Australia. The map shows the quite exceptional extent of their migrations, which cover some 12,000 miles at the minimum. One is tempted to say 'as the crow flies' but this is not strictly accurate since the terns do not fly in a straight line, so that the real distance travelled is perhaps even greater.

The common tern has an orange-red beak with a black tip and a tail that does not extend beyond its wingtips when at rest; its

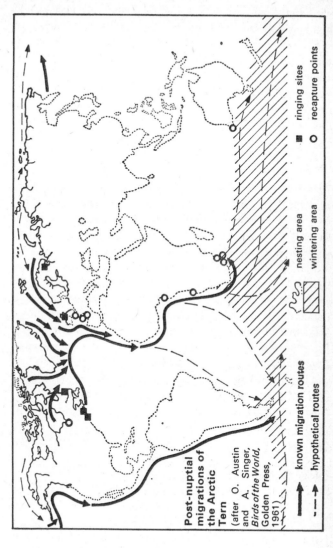

Post-nuptial migrations of the Arctic Tern

(after O. Austin and A. Singer, *Birds of the World*, Golden Press, 1961)

known migration routes

hypothetical routes

■ ringing sites
○ recapture points

nesting area
wintering area

Fig. 21

flight is capricious and twisting and it makes unceasing dives for fish. Without going far from the coast, it nevertheless reaches the Cape of Good Hope and from there it is sometimes carried by winds across to Australia. There one finds the Sandwich Tern (*Thalasseus sandvicensis*), a powerful tern with a long yellow-tipped beak, which can be identified from afar by its raucous and strident cry. Like the Arctic tern, it nests from the Arctic coasts of America, Europe and Siberia down to the island of Dumet (its most southerly nesting site). It then heads southwards and, having rounded the Cape of Good Hope, makes its way to Madagascar.

Skuas

These sombre birds, with their rapid and acrobatic flight, harry gulls and terns until they force them to disgorge the fish that they have caught. Nesting in the Arctic, the skuas spend the rest of the year flying over the seas. They are magnificent in the air and seem to be at ease even in the strongest storms.

The Arctic Skua (*Stercorarius parasiticus*), mainly a summer visitor to the British Isles, nests in the frozen wastes of the Arctic (America, Greenland, Iceland, Siberia) although some nest in the north of Scotland. After nesting the birds take to the high seas and winter in the southern hemisphere (Cape of Good Hope, Australia, Pacific and Atlantic coasts of South America). The Long-tailed Skua (*Stercorarius longicaudus*), which has two long and flexible central tail feathers (projecting five to eight inches, if they have not been broken), leaves in August and by October reaches the same winter quarters as the previous species.

The Pomarine Skua (*Stercorarius pomarinus*), which can be recognized by its blunt and partially twisted central tail feathers, has a quite different migration pattern. Thus, it has been recorded that some individuals have flown across the Atlantic to the Pacific and vice versa at about the level of the Panama Canal.

The Kittiwake (*Rissa tridactyla*) is about the size of the common gull but has black wingtips, a greenish-yellow beak, black feet and a more bounding flight (the immature birds have a characteristic black W-shape on the upper surfaces). This pelagic bird can be seen not far from our coasts and even in certain ports (Brest, for example). Except for certain British and Breton colonies, the kittiwake is a northern nester. Having devoted some months to reproduction (April to July), the colonies spread across the North Atlantic. Not uncommonly they follow a ship for the whole of its crossing of the Atlantic. In winter, flights of these birds cross the ocean in all directions as they are carried by the winds. If it becomes necessary to battle against the wind to avoid being blown over the land, the birds soon become tired and can lose up to 40 per cent of their body weight. In fact, most pelagic birds are unable to fly against the wind for long.

The order *Procellariformes* contains the truly great migrants of the high seas. Procella – the ocean – is the realm of the most fascinating of all sea birds, the albatrosses, petrels and puffins, whose very names conjure up the immensity of the lonely seas.

Better than any description are the series of maps showing the migrations of Wilson's Petrel (*Oceanites oceanicus*), prepared from the excellent research carried out by Brian Roberts. One most interesting result of this work was to show the strict relationship that exists between wind directions and the tracks of the migrants. This is now known to be fairly common in pelagic birds, which would seem logical.

The famous Storm Petrel (*Hydrobates pelagicus*), the smallest European sea bird and about the size of a swift, is a blackish bird with a distinct white rump, similar to Wilson's petrel but with a more flitting flight. Flying just above the waves, these petrels forage at the surface, for they often find themselves some thousands of miles from the nearest land. Sometimes, when high winds blow, numbers of birds gather together and this gives them their reputation as harbingers of storms, even as agents of nature's wild forces. The storm petrel lays a single egg in a rock crevice or in a burrow dug on an islet off Spain or in the Mediterranean, and then the birds fly to West Africa or the Red Sea. When they are on the move it is not unusual to see storm petrels off the French coast.

Leach's Petrel (*Oceanodroma leucorrhoa*) is difficult to distinguish from the preceding but it is slightly bigger and the tail is forked and not square, though it is seldom visible, however. It seems to be less inclined to follow ships. A northern nester, it undertakes immense journeys which bring it to the South Atlantic and into the Pacific.

The Fulmar Petrel (*Fulmarus glacialis*) is very similar to the common gull in size but is grey or uniformly yellowish and its

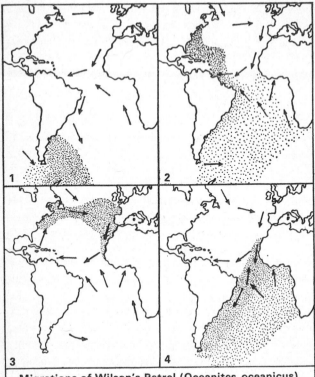

Migrations of Wilson's Petrel (Oceanites oceanicus)

▨ approximate distribution   ◄── direction of prevailing winds

1. In January the petrels are concentrated in their breeding area.
2. Then comes a rapid movement northwards. Arriving at the equator, birds on the western Atlantic side benefit from favourable winds, enabling them to travel up the American coasts, whereas birds off the African coast are held up by adverse winds (April).
3. In July the wind systems lead to the congregation of birds in the North Atlantic and their movement across to Europe.
4. The birds then descend to the south, and having battled to some extent against head winds (opposite the eastern tip of Brazil), they come back to their nesting area either by crossing calm regions or by taking advantage of the northeast winds that blow the length of South America (October).   (after Roberts, 1940, modified)

Fig. 22

extraordinarily elegant flight could not be confused with that of the former, which looks clumsy by comparison. Like all *Procellariiformes*, it has a beak with two nasal tubes used for excreting salt, for these sea birds must rid themselves of the great amount of salt they take in with their food or when they drink. It nests on the faces of cliffs in the Arctic and down as far as the Sept Isles (Brittany). During the winter it is the most abundant bird in the northern Atlantic, which it crosses with ease.

Puffins

The Puffin (*Fraterula arctica*) nests in large colonies on the off-shore islands of Scandinavia and Iceland, as well as Bermuda and the Azores. Like the petrels, they dig deep burrows in which they take refuge when they return from their flights. Very abundant along the French coast from July to October, they then make their way southwards and puffins marked in England have been recorded near South America in October. Their migrations remain more obscure than those of the Sooty Shearwater (*Puffinus griseus*), which has a sooty grey plumage. This bird nests in the southern hemisphere (Cape Horn, Chile, southern Australia) and winters in the North Atlantic (May) and passes up the coasts of Europe in August to September. It is along the American Pacific coasts, however, that it is most abundant.

The Greater Shearwater (*Puffinus gravis*) is a very remarkable bird. It breeds in only one place in the world, on the beautiful islands of Tristan da Cunha in the wilderness of the South Atlantic. From there the great shearwater travels northwards along the western borders of the Atlantic, quickly reaching Newfoundland and Greenland. Then, allowing itself to be carried along by prevailing westerly winds, it crosses the Atlantic and moves up the coasts of Europe in about August. In returning southwards once more, it flies over the eastern Atlantic and thus makes a gigantic circular tour by careful use of the prevailing winds. It is again the winds that aid the Short-tailed Shearwater (*Puffinus tenuirostris*) to make a similar circular tour of the Pacific after nesting in Australia. This enormous journey is, moreover, regulated with the precision of a clock, for in spite of the great distances travelled the birds arrive and depart from the nesting areas at remarkably regular times.

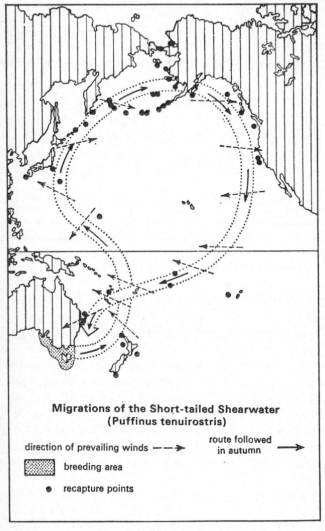

Migrations of the Short-tailed Shearwater (Puffinus tenuirostris)

direction of prevailing winds ‒ ‒ ‒➤

route followed in autumn ⟶

▨ breeding area

● recapture points

Fig. 23

Albatrosses

Lord of the oceans, the Wandering Albatross (*Diomedea exulans*), with its wingspan of eleven feet, is the largest of all ocean birds, and is capable of encircling the world every year in the southern latitudes where there is a continuous belt of ocean unbroken by land. To do this, it uses the westerly winds (the Roaring Forties) which, for two-thirds of the year, blow with storm force. That this albatross actually accomplishes such flights is attested by such cases as the Lysan Albatross (*Diomedea immutabilis*) which was released some 3,200 miles from the island of Midway in the Pacific and was recaptured there ten days later. Another case is that of a wandering albatross which, ringed on the Kergulen Islands (South Pacific), was recovered ten months later in Chile, having travelled at least 12,000 miles.

Shore birds

We have already seen that the plovers are capable of flying many thousands of miles over the open sea. From time to time one finds stray American birds reaching Europe, for example the Lesser Yellow-legs (*Tringa flavipes*), the Solitary Sandpiper (*Tringa solitaria*) and the American Stint (*Calidris minutilla*). The Siberian Pectoral Sandpiper (*Calidris acuminata*), which nests in Siberia, also strays as far as Tristan da Cunha.

Aside from these accidentals, there are the truly pelagic shore birds such as the Phalaropes. These rather curious birds (the female, for example, is more highly coloured than the male) are capable of settling on the sea, enabling them to feed in mid-ocean. They nest in the northern circumpolar regions and then descend southwards to winter off the western and southern coasts of Africa.

The birds mentioned hitherto are all regular long-distance migrants, but there is a small but important group of species which on occasions are also capable of astonishing flights. Although the occasions are isolated and exceptional, they nevertheless merit mention. In Europe, one can cite, for example, records during the last hundred years of the Black-browed Albatross (*Diomeda melanophris*), the Yellow-nosed Albatross (*D. chlororhunchos*) and the Cape Pigeon (*Daption capensis*) coming from the southern oceans, the magnificent Man-o'-War Bird (*Fregata manificans*) from the South Atlantic, Bonaparte's Gull (*Larus philadelphia*) from America, the Crested Auklet (*Aethia cristatella*) from the Pacific, and several others.

Penguins

We can end this review of sea-bird migrations with a short note on the Penguins. These flightless birds have, amongst their other peculiarities, the ability to migrate by swimming.

After nesting along the shores exposed to the full rigours of the south polar storms, each colony (adults as well as juveniles) moves northwards in the autumn. This journey is made under incredibly hard conditions. There are long marches across the wind-swept ice, alternating with long swims in places where the ice-floes have not yet formed (for this is the southern autumn). In spring, the penguins abandon their winter quarters and make the same tortuous journey back again. These powerful swimmers, whose aquabatics are wonderful to watch, are certainly less bothered by the swimming part of their journey than by the parts overland. This double method of migration by land and water is an interesting exception to general practice.

6 Migratory Behaviour in Birds

During migration, species of birds behave in various different ways. While in one species the birds may migrate in a group, in another they will fly singly; some may migrate at night, but others during the day. Let us examine this more closely.

First of all there is the question of the way in which the social behaviour of the birds is modified during migrations. Such modifications can lead to large assemblies of birds which would not normally congregate. Thus, there are many species which are solitary throughout the year but migrate in a group, for one of the principal factors in this solitariness – the need to defend a territory during nesting – does not obviously operate during flight. There are, however, some exceptions which will be discussed below.

What, then, are the species that migrate in groups? Theinemann recorded the passage over the Baltic ornithological station at Rossiten (Rybatschi) of flights of some 110,000 finches in 3 hours and 60,000 crows in 12 hours. Then there are the flocks of starlings, which bring together hundreds of thousands of individuals. The flights of shore birds (knots and black-tailed godwits for example) are frequently made up of thousands of birds, and the same is true of the flights of sand martins. S. Perry records that in 1939 he witnessed a remarkable assembly of geese:

In an hour and a half we could only watch in astonishment, for the numbers of geese were such that it was useless to try to count them. Even the thousands of geese seen on Loch Evan at the beginning of October were nothing to the hordes that stretched away into the distance. There were geese splashing in the shallow water, there were those that slept, heads under wings on the sand dunes, and others that ate the young reeds that bordered the shores for twelve miles. In our moored boat we made to take eight geese, which raised a flock of not less than 10,000. There followed a deafening noise of wings as flight after flight flew up in all directions, screaming loudly.

The Blue-headed Wagtail (*Motacilla flava flava*), linnets and ducks most often migrate in groups. It is noticeable that the species mentioned up to now have a strongly gregarious behaviour throughout the year. The sand martin nests in colonies, the blue-headed wagtail is rarely seen alone, and so on. But in birds which outside of the migratory phase are fiercely independent, there is usually no tendency to congregate during their travels. One even finds that birds brought together by force of circumstance during migration will try to keep as far from their fellows as they can. It would seem that they are still defending a territory much as they would on the ground. This is the case with certain shrikes, whose unsociability is well known.

There are, however, exceptions to this. Many places are famous for the very large numbers of birds that can be observed during migrations. For example, at the Hawk Mountain Sanctuary in Pennsylvania, some 10,000 birds of prey fly past daily – possibly more in the past but the United States in common with many other countries has managed to cause widespread sterility amongst birds of prey through the indiscriminate use of insecticides. Again, over the Bosphorus one can also see impressive numbers of birds of prey, sometimes with storks as well, and there are similar phenomena to a lesser degree over the Alps (at

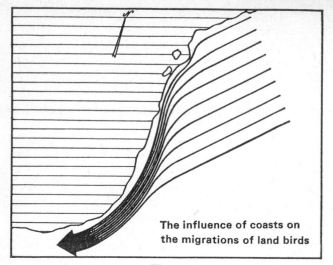

The influence of coasts on the migrations of land birds

Fig. 24

Brotelet) and in the Pyrenees. The explanations for such congregations of otherwise solitary birds are fairly clear. On the whole, birds of prey are not gregarious, the golden eagle, the peregrine falcon and the goshawk, for example, having strict hunting territories which they defend against intruders. But in the course of their travels there may be a local narrowing of the stream of migrants (a kind of migratory bottle-neck) in which they are temporarily concentrated. Cols between mountain peaks have this effect, while the case of the Bosphorus is explained by the fact that the birds that soar and glide, such as eagles, storks and the honey buzzard, avoid crossing the open sea at all costs and thus find themselves confined to the narrow bridge that joins Europe to Asia (see map, page 48).

Another phenomenon that brings together large numbers of migrants that do not otherwise travel in groups is the existence of assembly points and stop-overs. For many birds a migration route is not just a single flight from one place to the other but is made up of a series of stages between particularly favourable spots along the way. The case of water birds is a good example and in the migrations of ducks one sees a series of leaps from one water body to the next. There are also the great reed-beds that act as dormitories for swallows.

One can easily see that such especially favoured points must attract a great number of migrants, and at these places one finds all the various species that favour such habitats, although naturally not all of them migrate together. At large stop-over points for shore birds, where a number of species congregate, one finds that at the time of migration there is a very strict segregation: for instance, knots will not mix with black-tailed godwits, nor plovers with the various sandpipers. This is as expected, for the differences in migratory behaviour, especially speed, soon break up the heterogeneous groups of species into smaller flights of the same species.

Mixed migratory flocks can, however, occur when individuals of one species are swamped by a great number of individuals of another species. They are then forced to join the others, probably an instinctive urge to imitate, particularly in the case of shore birds and ducks. Birds with similar flying habits, such as rooks and jackdaws, or chaffinches and bramblings, sometimes mix on their migratory flights. There is also the case of the corncrake that flies with a small flock of quails (see page 64).

Diurnal and nocturnal migrants

Many representatives of these two groups have already been mentioned. Among the fairly strict day-time migrants are the swallows, the day-flying birds of prey, linnets, chaffinches, the brambling, and so on. Of nocturnal migrants the most characteristic perhaps are the warblers, the thrushes, the robin, geese and so forth. Others, such as the herons and the shore birds, are more independent in this respect. Birds that migrate in the day are generally those that can feed during the flight, such as martins and birds of prey.

It is among the nocturnal migrants that the migration itself most upsets normal routine. The migratory rhythms are very distinctly superimposed on top of their normal rhythm. Thus, when the sun goes down the birds cease their activities and prepare for the night's rest, as at other times of the year. But a few moments later the migratory urge seems to make itself felt and the birds rise and take to the air. It is probable that the birds fly throughout the night, for landing could prove dangerous in pitch dark, but from the cries of migrating birds and observations

The influence of contour on migration routes

In the British Isles migrants tend to avoid land over 600 feet.

Fig. 25

of migrants against the disc of the moon (see pages 42–3) there appears to be a maximum of flight activity a little before midnight.

7 Characteristics of migratory flight

Flight formation

While the majority of birds do not take up any definite flight formation, and one has only to think of a flock of starlings or swallows, there are others that adopt a well-defined pattern.

The most famous is undoubtedly the V-formation of ducks and geese, but typical also of pelicans, cranes and some shore birds (golden plover, for example). Various reasons have been advanced to account for this type of formation. Some claim that each bird leaves a 'wake' behind it and that the birds behind can rest one wing by taking advantage of the eddies thus formed. Others say that the turbulence from wing beats upsets a bird that follows it. Again, the V-formation provides an equal spacing between birds (by means of which they may then be able to judge distances better) and such a strict formation also ensures that all birds fly at the same speed.

There may also be a social factor involved. In many animal species there is a definite hierarchy (the peck-order of hens) and this may become translated into spatial terms in the V-formation. Large flocks of geese and shore birds can also form a vast solid triangle. Cranes, on the other hand, often fly in single file while rooks often form an elongated crescent.

Migrants which fly close to one another in a group are known to have a strong discipline on occasions and the cavorting of a flock of such shore birds as plovers shows a cohesion that recalls that of shoaling fishes, sudden changes of direction being carried out with superb synchronization. The formation of compact flights, which I have also seen in certain pelagic birds such as the Sabine's Gull (*Xema sabini*) – whose migrations are poorly known – must also be correlated with aerodynamic factors.

Speed of flight

There are two ways of looking at migratory flight speeds. First, there is the actual speed of the bird as it migrates, and second there is the average speed calculated for the whole of the migration, or at least for the main stages. Birds generally fly faster during migration than in normal flight.

Birds flying in a determined manner and on a straight course, as, for example, starlings, normally maintain speeds of about 30 mph, accelerating to 45 mph in parts of their flight. The yellow wagtail migrates at 30 mph, which is 6 mph faster than its general average flying speed. The situation is different with hunting birds, which attain tremendous top speeds. The peregrine (*Falco peregrinus*), for example, reaches 180 mph when 'stooping', but this only occurs during hunting and its migratory speed is about 30–40 mph.

Quite often the speeds of birds have been exaggerated, especially in ducks, and errors of this sort can easily arise if the observer does not ensure that it is the same flight of birds at the end as at the beginning. The following are some indications of the speeds flown by certain birds:

Mallards	46–60 mph migrating
Swans (Bewick's Swan)	39–42 mph normal flight
Chaffinches	21–29 mph normal flight
Swallows	20–46 mph migrating and normal
Quails	57 mph migrating
Cranes	22–36 mph migrating

These figures, which are from studies made by Richard Meinertzhagen, must be considered as approximate since a bird's speed will be greatly influenced by wind speeds. The fastest birds are the swifts and some of the ducks and geese, with speeds of up to 60–70 mph.

Turning now to average speeds for the whole migratory journey, these are of course much lower than the speeds measured at any particular instant. Migrants only devote a part of their time to actual flying (during which they do not feed but draw extensively on their reserves, chiefly fat accumulated prior to the migration), and most species make a series of stop-overs of varying duration during which they can feed and rest, so that when the migration is considered *in toto* the average daily run is perhaps not more than 30–120 miles. To make up for the rest period (and the ring ouzel, for example, may rest for five or six days at a stretch) there follows a flight which is made at a good speed. Thus, the Mediterranean or the Gulf of Mexico must be crossed in one go, although it now seems that birds can cross the Sahara in stages. There is a case of a Turnstone (*Arenaria interpres*) which was recaptured twenty-eight hours after being ringed some 510 miles from its place of capture.

In general, the spring migration is more rapid than that in the autumn. The Wood Warbler (*Phylloscopus sibilatrix*), for example, takes sixty hours to reach Africa from France, but only thirty to come back. One presumes that it is more worn out after the pre-nuptial than the post-nuptial migration, and that on its return it is more sensitive to the cold and rain that it finds on its arrival in Europe. Late frosts do, in fact, cause mass deaths among the chiffchaffs arriving in Europe from March onwards.

Altitude of flights

Certain birds fly at great heights. Storks have been recorded at 15,000 feet, geese and certain shore birds at 9,000 feet, swifts at 6,000 feet, and there are others that just skim over the ground or waves. There are circumstances, however, that sometimes require birds to go to great heights and certain species are quite capable of crossing such barriers as the Himalayas or the Andes at heights of over 20,000 feet. The record seems to be for geese flying over Dehra Dun in India at a height of 29,500 feet. In view of the physiology of the birds, this is an astonishing feat.

8 Influence of external conditions

Atmospheric conditions

Wind, temperature, the state of the sky – these are among the most important factors affecting migration. We have seen in the case of the swift and the swallow two very different types of migratory behaviour in relation to temperature. While the swift (and also the willow warbler, *Phylloscopus trochilus*) have very definite dates of arrival dictated by their 'internal clock' and the Snipe (*Gallinago gallinago*), the swallow, the Lapwing (*Vanellus vanellus*) respond much more to climatic factors. Such factors are, moreover, closely linked with the feeding of migrant birds, and this is especially true for birds feeding on insects; the birds time their migration to coincide with insect emergence, and the latter is naturally closely linked to such climatic factors as temperatures and rainfall. An example already mentioned is the standard-wing nightjar (see page 92) which moves north and south with the rain belt.

Winds, resulting from the general barometric situation, are also important in their effect on migrations. Quite often birds that are prevented from migrating because of adverse winds will wait for a change in the weather before attempting to proceed. It has been shown that Woodcocks (*Scolopax rusticola*) reach Hungary by means of the south-west winds, the latter arising from areas of low pressure over the British Isles. The influx of birds is greatly augmented at the time of the spring migration whenever this particular barometric pattern occurs.

The state of atmospheric electricity also seems to play its part and in general thundery weather seems to increase migratory movements, as Lars von Haartmann showed in a study of curlews. Birds, like many other animals, seem to be sensitive to stimuli which are barely if at all perceived by man (see for an example the section dealing with smell in fishes, page 152). It has been suggested that terrestrial magnetism influences migration in birds so that certain species move twice as far in places where magnetic anomalies occur.

9 The Physiology of migrant birds

The physiological changes that occur in migratory birds are complex and the exact relationship between these and migration has given rise to considerable controversy. This is particularly true in studies of the physiological stimulus to migration. For while it is recognized that birds, like other organisms, must be considered as systems comprising harmoniously co-ordinated parts, and that in the hormone system each part reacts on all the others, yet a number of research workers have assumed that one or other gland must dominate the rest as far as migratory behaviour is concerned. Thus, the migratory impulse has been attributed to the sexual glands by some, or to the thyroid by others. The relationships between the activity of these glands and migratory behaviour are real enough, but too often the relationship has been dogmatically interpreted as cause and effect, whereas it might well be that both gland and migratory behaviour are the effect of yet another cause; i.e. some third factor which influences the other two.

Quite often, the migratory and glandular activities are cyclic, so that one is in effect observing the coincidence of two such cycles. One should not go too far in this interpretation, however, for one cannot deny that through an internal rhythm and through the effect on it of external cyclic events, the endocrine organs can thus play a dominant role in preparing the animal for migration. Brought in this way to the final stages of readiness, the animal could then respond either to an external or an internal stimulus to trigger off the migration.

There are two extremes in migratory behaviour, between which are numerous intermediates. At one extreme, the migratory activity is a reflection of internal activity. The most striking example is found in birds which are raised from the egg under completely uniform conditions, with all seasonal variations eliminated, and yet show a cyclic migratory activity. The other extreme concerns birds that show a strong sensitivity to external conditions (temperature, day length, etc.), mainly through the intermediacy of the pituitary.* It is no doubt because birds vary in the degree to which they react to internal and external factors that experiments designed to show the relationship between hormonal and migratory activity so often produce conflicting results and imply that it is not possible to fit all types of behaviour into a unified pattern.

Internal rhythms

Apart from examples that will be mentioned shortly (i.e. birds which are reared in uniform conditions), the cases of the swift and

*The pituitary body, which lies in the lower part of the brain, is one of the most important control points in higher vertebrates. The anterior part, known as the hypophysis, secretes a number of hormones such as progynon (which stimulates the follicles of the ovary), androgen (producing growth of cells in the testes) and TSH (which affects the thyroid and thus influences such important changes as the storage or consumption of fat). The posterior pituitary controls, amongst other things, water excretion and pigmentation. In the following discussion, we shall refer merely to the 'pituitary gland', i.e. principally the anterior pituitary but including also the hypothalamus or stalk which connects the gland to the brain.

the golden oriole are often cited as examples of birds that depart before even a hint of the rigours of winter have come. Physiologists have suspected that, while the pituitary gland may be highly sensitive to external conditions, it can in other cases show a remarkably autonomous cyclic activity and in this way play an important role in the regulation of the 'internal clock' in migrant species. An apparently conclusive case is the fact that certain birds that pass from one hemisphere to the other maintain their own rhythm even though this is out of step with that of the local birds. This will recur for many generations, which shows that the extraordinary clock mechanism is written deeply into the hereditary complex of the bird.

Similar internal rhythms occur in some tropical birds living in an environment almost devoid of seasonal variation. This is so pronounced that in extreme cases, such as the Sooty Tern (*Sterna fuscata*), a species that breeds on Ascension Island, the reproductive cycle is less than a whole year (280 days in this case).

An interesing case concerns five subspecies of blue-headed wagtail (*Moticilla flava*) which winter in Africa. The physiological changes which bring the birds to a pre-migratory state (i.e. ready to migrate) do not occur simultaneously in the five subspecies, and the differences in time are directly related to the latitude of the place where nesting occurs after the spring migration. The most precocious subspecies is the one that nests on the periphery of the Mediterranean, while the last to reach a pre-migratory state is the one that flies as far as Scandinavia to nest and where the conditions favourable to nesting occur later. In Africa, however, and prior to this spring migration, the climatic conditions are quite uniform, so that one must assume that the differences in migratory behaviour are linked to the internal clock.

Influence of external rhythms

Among the many repetitive events and cycles occurring in the natural environment, there are certain ones that seem to have overriding importance in their influence on birds. Light and temperature are two that are both easily experimented on and quite clearly appreciated by birds.

Light and bird behaviour are difficult to separate since, outside the actual period of migration, the activities of birds are so strongly affected by light. Nevertheless, certain experiments carried out by Professor Jacques Benoit on Pekin ducks have demonstrated that the duration of light periods (i.e. day length) can act specifically on the pituitary hypophysis and hypothalamus and thus influence the development of the sexual organs and migration. The pituitary body acts as an intermediate between the environment and the internal organs of the body, so that light falling on the retina of the eye stimulates the optic nerve, which in turn activates the hypothalamus or pituitary stalk, leading to the stimulation of the hypophysis itself and the production of hormones that activate the gonads. Professor Benoit did, in fact, produce such a chain of events when the ducks were blinded and the pituitary stimulated by shining a strong light through the roof of the skull (which is thin in birds) or by projecting the light through by means of a quartz rod. As these experiments showed, the pituitary is a gland of the greatest importance in the organization of bodily activities, one of which is the process that leads up to migratory behaviour.

Let us examine the effects of changes in the length of day and night, referred to as photoperiodism. In the plant kingdom some famous experiments have shown that flowering in certain plants (sugar cane, for example, or some species of tobacco) is conditioned by the lengthening of the nights in the autumn, and that if one interrupts these long nights (by shining lights on the plants) flowering does not take place. It is noticeable that a similar phenomenon occurs in birds. The spring migration has been shown to have been conditioned by light periods during the preceding winter. The long nights (of about fifteen hours) are necessary as a first phase in the slow but profound changes that lead up to the pre-migratory state. Why 'long night' rather than 'short day', which on the face of it would seem to come to the same thing? The reason is that the most important factor (as in the case of the plants mentioned above) is the length of the nights, i.e. the dark period.

This first phase, of lengthening nights, is followed by a second phase, characterized by increase in day length at the end of which the full pre-migratory state is reached and the bird is ready for flight. But if the first phase is inhibited by artificially shortening the nights, the second phase is not influenced by lengthening the duration of the nights. The explanation seems to be that the long nights of autumn are a passive factor necessary for the creation of the first phase, and this passive factor is upset by the active factor of long days. On the other hand, the active factor of the second phase (long days) appears to be uninfluenced by the passive factor of long nights.

In the case of the autumn migration, the situation is much more complex since many birds move to the other hemisphere and find themselves suddenly faced with days that are growing longer rather than shorter. This one would logically expect to inhibit the first pre-migratory phase destined to prepare the birds for the spring migration. To account for this it has been assumed that the pituitary enters a refractory period after migration and breeding and that for a time it is insensitive to variations in day length.

Certain authors have supposed that it is not so much light *per se* that acts on the bird, but its concomitant – wakefulness. They have pointed out that one can obtain results similar to those got by the direct action of light merely by regulating the periods of activity of the bird. In such experiments the animal is exposed to a low level of light which is insufficient to produce the usual responses and at the same time it is forced off its perch by a revolving shaft. The birds can then be kept awake for the desired periods. In the final count, it would appear that light and activity are closely linked and are not in any way in opposition to each other but rather act together (the one perhaps even reinforcing the effect of the other).

The other factor which has been strongly emphasized in studies of the influence of external stimuli on birds is temperature. In the course of his experiments on photoperiod, Benoit showed that gonadal development that was stimulated by light was completely independent of the ambient temperature. Thus, the responses to variations in photoperiod were the same in batches of ducks held at $4°C$ as in other batches held at $20°C$. Benoit's experiments centred on the development of the gonads under the influence of light periods, but this development, to-gether with the phases of migration, is controlled by the pituitary and their cycles coincide so closely that for a long time it was assumed that the gonads directly affect migration. Temperature, however, certainly plays a part, often as the final trigger setting off migration itself.

Warblers, for example, when they are fully ready for flight, show increased activity when the temperature drops. According to some research workers, cold stimulates certain hormone secretions from the thyroid gland and thus affects the metabolism of the body fats which, as shown below, is a physiological process closely connected with migratory activity.

This reaction to cold is most marked in birds that maintain a high body temperature and as a result must feed abundantly. The thyroid plays a major role in the regulation of basal metabolism, for hyperthyroidism caused by experimental or natural means can raise the metabolic rate by some 40 per cent, which is a considerable increase. In spring, however, one obviously cannot invoke cold conditions as the stimulus to thyroid activity, and the release of hormones must therefore be part of an internal cycle in the gland's activity. Nevertheless, if a late cold snap takes the birds by surprise during their pre-nuptial migration, then the increased activity of the thyroid resulting from this leads to a resorption of fat reserves and a stop to the migration.

Turning to the equatorial regions, it is immediately apparent that, since day length is virtually constant, it is not photoperiod but the alternation of dry and rainy seasons that must influence the cycles of the birds.

It may be objected that in very many of the experiments performed on birds the animals are captive and kept in cages, and that this must influence the results. The answer to this is that it has been repeatedly shown that the annual cycles (hormones, gonads, general activity) in captive birds are identical to those of birds in the wild.

The Pre-migratory state

The various physiological events that occur in birds in the preparatory period before migration culminate in a state which can be termed 'pre-migratory'; in this condition the bird is susceptible to the final trigger that sets off migration. In some cases the trigger is temperature, in others it may be lack of food, and so on. It may also be the internal alarm clock ringing, so to speak.

At this stage the bird has built up reserves (chiefly fatty tissues) on which it can draw during the long flight. Such reserves are especially abundant in diurnal seed-eating species which cannot feed on the way. The fat accumulation naturally leads to an increase in body weight and this is very obvious when one compares the weights of migratory and sedentary members of the same species (i.e. of partial migrants, see page 45). Carefully compiled figures indicate that in migrant birds the weight increases steadily from December to April, while in sedentary birds there is mostly a tendency for weight to decrease. Considerable differences occur between species. Thus, in the United States it has been found that the percentage of fat in total body weight was only a few per cent in buntings that did not cross the Gulf of Mexico but rose to as much as 50 per cent in the long-distance migrants.

The appetites of birds are not always greatest immediately before the two migrations, for if one provides different species with a superabundance of food one finds that the complete satisfaction of alimentary needs actually calms down certain small migrants in which the migratory urge is not very strong, whereas it does not hinder the departure of long-distance migrants.

It is chiefly the hormones of the thyroid, under the strict control of the pituitary, that regulate weight in birds. The anterior pituitary (i.e. the hypophysis) secretes a hormone called thyrostimuline which, as its name suggests, promotes thyroid secretion. The hormones from the thyroid, however, have an antagonistic action on thyrostimuline, and in this way a natural balance of forces is achieved. The thyroid hormones have the effect of increasing basal metabolism and oxidation processes and in particular they promote the breakdown of fats, thus allowing the bird to use up its fatty reserves quickly. There is a striking correlation between the thyroid cycle and that of migration (see Fig. 26). Migratory activity is greatest during the resorption period (solid line on graph) when the hormone is liberated into the body. During the secretion period that follows (broken line), hormone is produced, but it accumulates in the follicles of the gland and is not passed into the bird's system (hatched line). It can also be noted that moulting is closely influenced by thyroid activity, and that the removal of the thyroid gland can inhibit moulting.

Another phenomenon is that during the migratory period

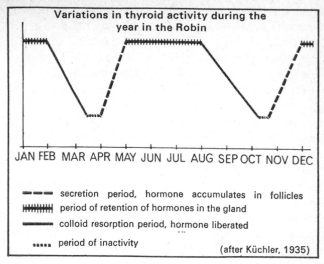

Fig. 26

birds can replace, that is to say actually build up, fat reserves more readily than at other times of the year even though considerable energy is being used in the migration itself. Thus, both the build-up and the breakdown of fats are heightened and one can say that the whole complex of metabolism is at this time designed to cope with the mobilization and expenditure of a great amount of energy.

The pre-migratory state is also characterized by the developmental state of the gonads. Experiments on the relationship between migration and the sexual cycle have not, however, produced decisive results. This research has had great descriptive value, but the interpretations placed on the results by their authors cannot be accepted at the present time. The interpretations have tried to establish a relation of cause and effect between sexual and migratory events. In reality it would seem that there is merely a coinciding of the two cycles, each under the ultimate control of the pituitary. This is certainly the case as far as the gonads are concerned, and probably so also with the migratory urge.

Apart from their specific role in reproduction, the sexual glands also exercise an important influence on the metabolism of the animal. The androgens or male hormones, for example, stimulate protein build-up and play a major role in the bio-chemistry of muscle, assisting in the reconstruction of high energy molecules required for muscle contraction.

The first major work on the relationship between the gonads and migration was that of William Rowen, a physiologist who worked in Canada on the Slate-coloured Junco (*Junco hyemalis*), a North American bird that has occasionally been recorded from Ireland, Iceland and Italy. Rowen found that at the time of migration the gonads were in an intermediate stage of development. Birds which showed the maximum or minimum development of the gonads exhibited non-migratory activity.* This is not in any way surprising and merely reflects harmony between the two cycles. In springtime, which for almost all birds is the time of reproduction, the gonads have obviously reached their peak of development and the very act of nesting implies the cessation of large-scale movements by the birds. In winter, on the other hand, the gonads are in a state of minimum development, while the birds are in their winter quarters. However, Rowen then activated the gonads by means of altering day lengths (i.e. through the action of the pituitary, as discussed earlier) and through variations in the photoperiod was able to stimulate no less than three maxima in gonadial development in a single year (the corresponding minima were easily achieved by shortening the light period by six minutes a day). We have already seen that the sensitivity of the animal to differences in day length is variable for an increase in day length can suppress the autumn migration, while in spring the same increase can have a stimulating effect. Basing his theory on incontestable facts, Rowen arrived at the conclusion that the gonad state has a direct effect on migration. His further studies, on the American Crow (*Corvus brachyrhynchos*), produced certain exceptions which detracted from the simplicity of his earlier conclusions and led him to state that only the spring migration was dependent on the gonads. Thus, the migration of castrated birds clearly showed that the gonads could not stimulate migration. Further experiments showed that cyclic migratory activity is identical in both castrated and normal birds, and that the development of the gonads must accompany rather than initiate migration.

*The length of the testes in these Juncos varies from 0·5 millimetres in November to 7·0 millimetres in May.

The study of partial migrants can make a significant contribution to the problem of the physiology of the pre-migratory state since one finds differences between sedentary and migratory individuals of the same species (differences in weight have already been referred to). It is, however, necessary to distinguish two types of partial migrant. In the first group, of which the herons are a good example, individuals of the same population in a given species show different migratory patterns, ranging from the completely sedentary to the long-distance migrants. Thus, in one and the same colony of herons one will find birds that will cross the Atlantic and others that will scarcely move at all. Again, in numerous passerine species the migratory urge seems to depend on the age or sex of the individuals concerned. Certain birds which only reach sexual maturity in the second year, such as gannets, sometimes remain the whole year in the wintering area of their parents. The albatrosses only breed every two years and during the interval display literally a global nomadic behaviour. We have also seen that there are differences in migratory behaviour correlated with sex and that the Latin name *coelebs* (bachelor) for the chaffinch is a fair description, the males as in other passerines being much less migratory than the females. Some workers have concluded from this that the male hormones actually oppose the migratory urge and lead to the sexual rejuvenation sometimes seen in autumn (certain birds, such as the chiffchaff, begin to sing once more and in the plant kingdom there is the example of the hawthorn which can begin to flower if there is a warm spell in autumn).

The second type of partial migrant concerns species in which the individuals have different migration patterns depending on the population to which they belong. We have seen, for example, how partial migrants in Europe are more migratory in the north and east than elsewhere. Christian Erard showed that in France the robins of the south-west are more sedentary than those in the north-east.

Another aspect of the physiology of migration is the moult. All birds have a post-nuptial moult and generally there is a second moult before the autumn migration (occasionally taking place at the winter quarters). The moult coincides with the release of a large quantity of thyroid hormones and moulting can be stimulated artificially by stimulating the thyroid. The 'moult migration' is a particular aspect of this phenomenon (see page 53). As already mentioned, certain birds such as ducks, waders and rails lose their flight feathers completely at the time of moulting and, being vulnerable and needing peace and security at this time, they must make a preliminary migration to a select area.

In conclusion, it can be said that migration sets in motion a whole physiological metamorphosis under the influence of the autonomous hormone cycles and the pattern of external stimuli, the two often being combined. In most cases the imprint of external conditions leads to internal changes which are reflected in the activity of the pituitary complex and this body adds its own internal rhythms.

Albert Wolfson, in a series of excellent studies on the Oregon Junco (*Junco oreganus*), deserves the credit for showing that simultaneous changes in the thyroid or the gonads are, like the migratory phenomenon, under the control of the pituitary hypophysis and that concordance between the changes does not necessarily imply a cause and effect relationship between the thyroid or gonads and migration.

10 Orientation of birds

For human beings, the ability of birds to orientate is both astounding and difficult to appreciate fully. While all cases of orientation and navigation are fascinating, the abilities of the swift are perhaps the most extraordinary, and particularly the way in which the birds return with clock-like precision at the same time and to the same place, perhaps for several consecutive years, after spending each winter in Africa. Another equally striking example is the golden plover of the Pacific which, without sextant or chronometer, takes a fix on the Hawaiian islands from Alaska and sets off on a flight which, were it to deviate only 1°, would mean an error of about one mile after only sixty miles of flight, whereas the birds cover some 2,065 miles on their journey. Also baffling is the 'Kaspar Hauser' type of bird. Kaspar Hauser was a mysterious young man who appeared in the streets of Nuremberg in May 1828, dressed in peasant garb and bewildered by his surroundings. According to the letter in his possession, he had been given into the custody of a peasant who was instructed to educate the boy sixteen years previously but keep him in close confinement (one theory was that he was the heir to the Grand Duke Charles of Baden, kidnapped by the minions of the latter's morganatic wife). Hauser claimed to have been kept in a tiny room or cell and to be ignorant of the customs and ways of men. Kaspar Hauser birds are those that have been raised in isolation so that one can determine the part played by heredity in their subsequent behaviour. In autumn, such birds resolutely face to the south-west when they are first shown the night sky.

The first systematic studies on orientation in birds were made possible by the 'homing instinct' exhibited by so many species. Although perhaps a little artificial, this method has yielded useful results. Birds are caught at a time when they show an attachment to their territory (especially during the nesting season). After taking them to some spot, one releases them and records the percentage of returns. The distance can be varied, and the direction, as well as the method of transporting them and one can then study the influence of climatic and other factors on their ability to find their way home.

These experiments have shown a wide variation in ability to home, some species being better endowed than others. This led Donald Griffin to distinguish three types of homing behaviour.

The first type is the use of visual landmarks, the birds methodically exploring the terrain in which they are released until they pick up some familiar feature, whereupon they quickly find their way back to the nest. Such birds possess a highly developed visual memory, as experiments with pigeons have shown. Domestic pigeons have been trained to peck at a certain point on an aerial photograph (through a system of rewards), and four years later the birds were still able to respond to this training when placed on the aerial photograph. Birds' eyes have a power of resolution two to three times greater than ours, enabling them to pick up very fine details. If a bird only uses the Type 1 homing behaviour, homing can only succeed if the point of release is not too far away. If the birds are transported 500 miles from their nesting territory, it is only by good fortune that they find their way back as a result of long exploratory flights. Usually the area known to a bird is its hunting or feeding territory, which in the titmice probably does not reach more than five or six miles from the nest, but which can cover hundreds of square miles in the case of birds of prey. Released in their feeding territory, the birds soon make their return; release them outside it and much fewer return. Trying this a second time, the visual memory comes into play and the bird, no longer requiring tedious exploratory flights, will return much more quickly. Homing birds have also been followed by aeroplane. Donald Griffin and Raymond Hock followed Gannets (*Morus bassanus*) in this way. The birds had been released some 200 miles inland from their coastal nesting sites and although two-thirds of the birds had found their way back, this was only after long, meandering flights, showing that they did not possess the directional faculties of long-distance migrants.

The latter also use the method of visual landmarks in the final parts of their journeys which enables them, after finding the general area of their territory by other means (see below), to nest year after year in the same place (swifts, wrynecks, swallows, etc). There are even some rather rare cases of migrants that are faithful to their wintering place. A wagtail, for example, is known to have returned for several years to the same garden near Bombay.

The second type of homing behaviour is shown by birds that are capable of choosing their flight direction and holding to it for the rest of their journey. How do they decide what direction to take? They appear to choose their normal direction of migration even though they are now in a different place, i.e. the point at which they are released. If, for example, birds which normally fly to the north-east to reach latitude 45°N are released at that latitude they will immediately start flying north-east *regardless of their longitude*. If they are released along their normal migration route they will easily reach their nesting area, but if one releases them on the same parallel but further to the west they will maintain their correct course but pass by to the west of their area. Werner Rüppell showed how Hooded Crows (*Corvus cornix*), caught at the ornithological station at Rossiten on the Baltic and released 600 miles to the west, were for the most part recaptured to the west of the area in which they would normally have nested and, more interesting, that their wintering area was also shifted to the west. Thus the birds clearly main-

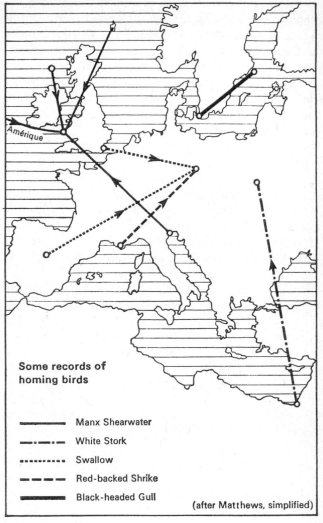

Amérique

Some records of homing birds

————————	Manx Shearwater
—·—·—·—·—	White Stork
············	Swallow
— — — — —	Red-backed Shrike
▬▬▬▬▬	Black-headed Gull

(after Matthews, simplified)

Fig. 27

tained their normal north-east/south-west axis of migration but both routes had been pushed westwards.

The third type of homing behaviour shows the highest degree of orientation. Released at one point, the birds immediately take stock of it, compare its position with that of the nest, decide on the direction and fly off. This happens even if the birds are completely out of their element in a country right off their migration routes and where they have never been before. In one famous example, a lysan albatross returned to its nesting area on Midway Island in the middle of the Pacific, having flown 3,200 miles from Widby Island (Washington, USA) in just over ten days. This is a perfect case of the third type of homing, for the albatross clearly could not rely on any landmarks over the vast expanse of the Pacific. An even more remarkable case was that of a Manx Shearwater (*Puffinus puffinus*) that returned from Boston (Massachusetts) to Skokholm off the coast of Wales in twelve and a half days after a journey of 3,050 miles across the Atlantic, an area scarcely frequented by these birds. Another bird returned from Venice in fourteen days after crossing the entire continent.

The percentage of successful birds varies greatly, being highest in those species with a strong migratory behaviour. Thus the

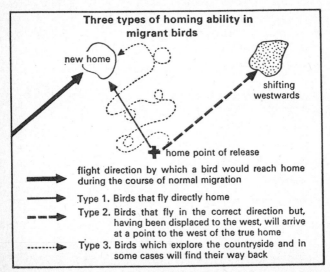

Fig. 28

lesser black-backed gull is more migratory than the herring gull and more often reaches 'home'. Great migrants such as the swift have the highest percentage returns. In one case, seven out of nine alpine swifts were recaptured at their nests after being displaced some 900 miles; one made the journey in three days.

What part does heredity play in all this? Two experiments suggest that instinctive, i.e. genetically inherited, behaviour patterns play a part in navigation. The first was carried out by Ernst Schüz and it is highly significant. Schüz caught first-year European storks and released them only after the departure of the adults at a time when they normally make their south-west autumn migration (see map, page 50). The recaptures showed that, in spite of the fact that there were no adults to guide them, the birds unanimously headed south-west. This was a most striking experiment, for it showed that the birds had an innate and unlearned attraction for the African wintering area that they have occupied for thousands of years.

The case of starlings is a little different. These birds have a great aptitude for homing (seven out of eight birds, for example, returning after a flight of about 400 miles), but this behaviour differs in the different age groups. Birds that were shifted to the south-east of their normal migration route split into two lots. The adults, in full possession of their gift for orientation, found their wintering area by modifying their direction by 90°, whereas the juveniles behaved like the hooded crows mentioned earlier and sought their winter quarters (to which they were nonetheless faithful) to the south-east of its real position.

Such performances raise a torrent of questions, for it is at present still impossible to explain how the birds manage to orientate themselves. Numerous explanations have been advanced, some reasonable, others fantastic, testifying to the complexity of the problem. To test such theories, a number of subtle experiments have been devised, the results of which have been quite spectacular. We shall see, however, that the question is still not totally resolved and that a final answer is still some way off.

The basic proposition is this: birds are apparently able to identify their position, much as a sailor does with sextant and chronometer, and then to navigate by means of astronomical points of reference (the stars for nocturnal migrants and the sun

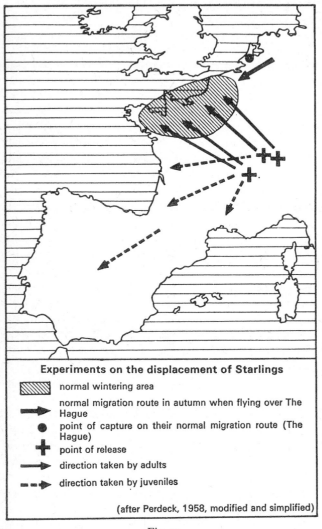

Experiments on the displacement of Starlings

▨ normal wintering area

➤ normal migration route in autumn when flying over The Hague

● point of capture on their normal migration route (The Hague)

✚ point of release

──➤ direction taken by adults

━ ━➤ direction taken by juveniles

(after Perdeck, 1958, modified and simplified)

Fig. 29

for those that travel by day). Let us examine the facts that support such a conclusion.

Nocturnal migrants

Dr Franz Sauer was studying the behaviour of nocturnal migrants and in working on warblers he noticed a curious thing; the birds would align themselves in the direction in which they were going to fly before they actually took off, as distinct from other birds that rise in the air, circle for a while and then finally head off in the right direction. This behaviour of the warblers was of the very greatest use in Sauer's experiments, for one needed only to enclose the birds in some sort of cage with a graduated circular perch in the middle, and then record the direction in which the birds face. After a moment or two of hesitation, the birds take up their positions on the perch and show every sign of being ready to fly off in that direction. If one rotates the perch, the birds will adjust their positions accordingly. The direction chosen by the birds was found by Sauer to be that which they normally take in the autumn migration, which for the Lesser Whitethroat (*Sylvia curruca*) is to the south-east. This experiment was carried out in fine weather, out of doors and with a clear starry sky.

Sauer found that on nights which were cloudy and the stars obscured the warblers were incapable of orientating and would wander round the perch. But if there was a break in the cloud cover and if one or two constellations were visible, the warblers would immediately align themselves facing south-east once more. This was confirmed by Frank Bellrose in some experiments carried out in 1958 using mallards marked on the foot with a luminous patch so that they could be followed at night. Bellrose's two graphs (see Figs. 30 and 31) clearly show that when the sky was covered the mallards flew off in all directions whereas when it was clear they unanimously headed northwards.

As a result of such experiments, Sauer decided that it was the stars that were the principal factor in navigation and to verify this rather bold assumption he devised a most ingenious experiment. He made the birds repeat the performance, but this time in a planetarium. He took the whitethroats, together with the equipment described above, to the Olbers Planetarium at the Naval School in Bremen, where he carried out a series of most

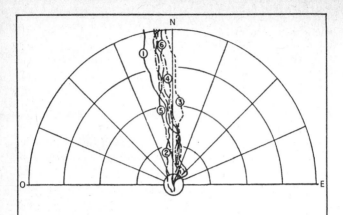

Trajectories of six Mallards within a radius of one mile from the point of release on a clear night with the stars visible.

(after Bellrose, 1958)

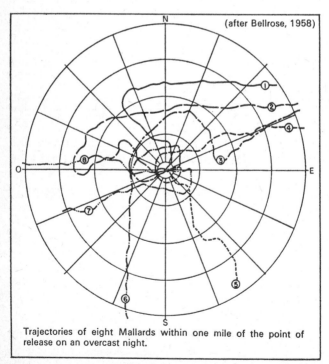

(after Bellrose, 1958)

Trajectories of eight Mallards within one mile of the point of release on an overcast night.

Fig. 30 (above) and 31

imaginative experiments. First of all he showed the birds a replica of the night sky as it actually was outside the planetarium, with the result that the birds correctly faced to the south-east and at least showed that they were able to respond quite normally to the artificial stars. This, however, proved nothing, for the birds could quite well have been responding to some other factor (magnetic or electric fields, cosmic rays, and so on). The decisive experiment was to see what would happen if the whole night sky were rotated 180°. The Pole Star, for example, would then be in the south and not the north as far as the birds were concerned. The result was quite remarkable. The birds actually followed this rotation and ended up facing to the real north-west (i.e. what appeared to be the south-east on the planetarium ceiling). Many such experiments were carried out, sometimes rotating the sky 45°, sometimes 90°, and so on. There was no doubt, the birds aligned themselves to face what appeared to be the south-east. All extraneous factors had been removed and the birds were clearly orientating with reference to the stars. If the stars were then extinguished, the birds were incapable of orientating and behaved as they did when the real night sky was clouded.

We shall deal further on with some elementary mathematics of celestial navigation, but one thing must be mentioned at this point. As is well known, the stars appear to move during the night and in our hemisphere they circle the Pole Star. Thus certain stars rise and others set and the night sky is never the same from moment to moment. To be able to orientate and navigate one must know, therefore, the exact time, for a star that is directly south-east at ten o'clock is not in the same position by eleven. This supposes that the whitethroats are capable of three highly complex abilities.

1. They must be able to recognize constellations.

2. They must have an internal clock by which they know the exact time.

3. When they see a particular night sky they must know at such a day and such an hour precisely where the south-east lies.

But this is not all. During these experiments one must try to construct artificially the kind of conditions that the bird will encounter during its migration. In the case of the lesser white-throats, the German populations during their autumn migration first of all migrate to the south-east, passing over Bulgaria,

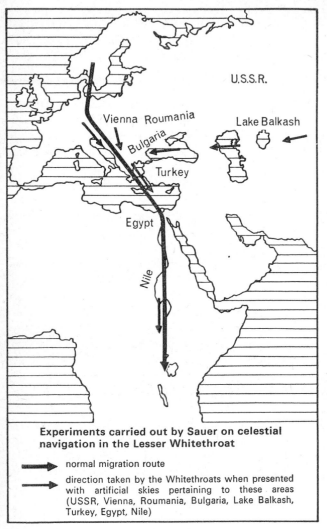

Experiments carried out by Sauer on celestial navigation in the Lesser Whitethroat

→ normal migration route

→ direction taken by the Whitethroats when presented with artificial skies pertaining to these areas (USSR, Vienna, Roumania, Bulgaria, Lake Balkash, Turkey, Egypt, Nile)

Fig. 32

Turkey, traversing a part of the Mediterranean near Cyprus and then changing course and heading due south up the Nile valley. Going south, of course, one finds new stars and constellations appearing, so that the night sky over Cairo is different from that over London. Therefore, the birds were shown the night sky as it appeared over Bremen and they orientated south-east. They were then progressively shown the skies over Czechoslovakia – they faced south-east, Hungary – they faced south-east, Bulgaria and Turkey – they faced south-east, and then all at once the sky over Cyprus, and without hesitation the birds faced to the south as if to head for the Nile.

The birds were not bothered by the fact that during this experiment a month's migration was compressed into three hours (though one wonders how this affected the internal clock). If the birds were then shown the night sky corresponding to the end of their journey (south of the Nile valley) the migratory behaviour ceased and the birds took up their normal rhythm of life and went to sleep as they would normally outside the migratory phase. One can, however, complicate their lives some-what. Show them at nine o'clock in the planetarium at Bremen a sky that would not normally be visible at Bremen until ten past two in the morning, and what do they do? In fact they are non-plussed. They appear to realize quite well that the correct time is nine o'clock and that at this time such a night sky should not be seen over Bremen. They draw the logical conclusion: it is not Bremen at all but Lake Balkash in Siberia, for that is the only place where such a sky could be seen at such a time. Here, one must remember the hour zones, for an advance of five hours and ten minutes is equivalent to a displacement of $77°E$, that is to say the degrees of longitude between Bremen and Lake Balkash (the earth makes a complete revolution of $360°$ in twenty-four hours and thus turns $15°$ an hour). Thus an observer on Lake Balkash sees a configuration of stars which is not visible in Bremen for another five hours and ten minutes.

We left the whitethroats believing that they were some distance to the west of their real position. What do they do? Continue to take the migration route to the south-east as in all previous experiments? On the contrary, the birds which have never been to Lake Balkash and have never in their lives been faced with such a problem, actually correct their positions and now face west as if to retain their normal migration route. They hold to this position right up to the night skies appropriate to the Russian frontier and then, when the Roumanian sky is seen, they face southwards and if one makes them 'arrive' in Bulgaria (solely by using the planetarium sky) they then recognize their normal migration route and turn to face south-east as they would do on their usual journey.

This relationship between the whitethroats and the stars seems quite incredible, but the facts are there, clear and unequivocal. The experiments can be repeated or modified. If the planetarium sky is, for example, altered by only one hour then the birds deduce that they are now over Venice and they face southwards until they encounter a celestial pattern that agrees in time and season with what one would expect had they rejoined their proper migration route. Each time the birds respond without hesitation. One can, of course, place them in an absurd situation for which there is no natural equivalent. One can show them a spring sky in autumn, in which case the birds are torn between their internal clock which tells them that it is autumn and the stars that argue that it is spring. Such a situation does not occur anywhere on the globe and the birds do not know how to respond and face desperately in all directions. This shows that the white-throats do not respond to some blind, mechanical urge which has no relation to actual events.

The last of these fascinating experiments concerns the 'Kaspar Hauser' whitethroats (see page 125). Showing them the autumn night sky such as would correctly be seen at Bremen, they hesitate a moment and then, certain of their decision and driven by some deep instinct, they turn and face to the south-east. And yet, some seconds before this they had never in their lives seen a night sky.

Diurnal migrants

Some experiments analogous to those carried out by Sauer were undertaken by one of the great experts on bird orientation, Gustav Kramer. These experiments established the role of the solar compass in the orientation of certain birds. The following is a review of this work.

For these experiments Kramer used the starling. The birds were placed in a circular enclosure in which there was nothing

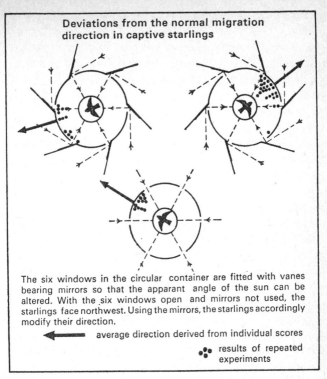

Deviations from the normal migration direction in captive starlings

The six windows in the circular container are fitted with vanes bearing mirrors so that the apparant angle of the sun can be altered. With the six windows open and mirrors not used, the starlings face northwest. Using the mirrors, the starlings accordingly modify their direction.

← average direction derived from individual scores

•• results of repeated experiments

Fig. 33

that they could use as a point of reference, for the slightest irregularity which might serve as a landmark could upset the experiment. This enclosure was lit by six windows which allowed the penetration of the sun's rays. Outside the windows were fixed vanes which could be swung and on these were attached mirrors so that by moving the vanes the apparent direction of the sun could be altered. The bottom of the cage was transparent so that an observer could record the position of the bird. This was noted every ten seconds and the results worked out from a statistical analysis of the experimental data.

1. All the windows were opened and the mirrors not used to modify the sun's direction. The starlings then orientated to the north-west at the time of the spring migration and showed migratory activity (a certain fidgeting). As already stated, the

migratory cycle is practically the same in both caged and wild birds.

2. When, by means of the mirrors, the direction of the light rays was altered, the starlings reacted accordingly, as the diagrams on page 137 indicate. This clearly showed that the factor concerned was the direction of light and not something else. This was confirmed in the case of electro-magnetic waves for it was found that placing a quantity of iron next to the cage made not the slightest difference to the birds although a considerable difference to a compass; also, an electro-magnetic beam had no effect.

3. In modifying the positions of the mirrors it was found that a direct view of the sun itself was not indispensable. Quite significant results were found when the birds were influenced by the reflection of a part of the sky close to the sun; in fact, an area of 45° either side of the sun was sufficient to activate the birds. (We shall see the same phenomenon in bees when dealing with the effect of polarization.)

4. If the light was diffuse and even, the birds were incapable of orientating. This corresponds with what we know of pigeons and other birds. If the sky is covered, then the released birds take the wrong direction but as soon as the sky clears a little they orientate correctly. If, however, the migrants fly in cloudy conditions this is because it is rather rare not to find sunny intervals at some time or other and in the meantime the migrants use geographical aids such as rivers, coastlines and mountains.

5. The birds evidently took the movement of the sun into account. Just as nocturnal migrants correct their flight path in relation to the rotation of the stars round the poles, so diurnal migrants must appreciate the movements of the sun and, with the aid of their internal clock, compensate at any time in the day and thus orientate in the right direction. Thus to fly on a steady north-west course they must use a moving object (the sun) to steer by. This was demonstrated in the following experiments. A bird was trained to take food from a receptacle at a precise time (a time when the sun always occupied a set position). The training ended, the birds were placed in an experimental cage in which were a number of identical feeding trays, and at the same hour of the day the bird, having compensated for the movement of the sun by means of its internal clock, selected the correct feed-

ing tray. By means of an artificial sun, one can do what has already been done often to plants, that is one can upset the internal clock by progressively modifying the periods of 'night' and 'day'. The subject then orientates incorrectly on being returned to natural conditions.

Many theories have been proposed to explain the way in which the famous solar compass works, for it is well known also in insects and fishes. Geoffrey Matthews proposed the following solution.

The birds, he believed, must be able, from relatively short observations of the sun, to deduce the remaining part of the complete arc traversed by the sun throughout the day, and especially its highest point (i.e. at noon). In other words, the birds must be able to appreciate the angle marked A on the diagram on page 140 which bears a relationship to the latitude of the place where the observation is made. This is what sailors refer to as taking the meridian. At the same time, in homing birds, the animal must deduce its longitude by measuring with its internal clock the time displacement between the real position of the sun and that which it would occupy were it to see it from its destination.

The details of this theory are debatable. Thus it is hard to imagine that a bird can see the sun for some ten seconds and from this extrapolate the sun's complete trajectory across the sky. To make such a measurement the bird would have to remain still, which it does not do. Again, the speed of this very precise calculation must be very great if birds are to react as fast as the previously cited experiments show. An error of 1° (which is extremely little) in estimating latitude would lead to an error of many miles at the journey's end. Finally, the greatest drawback to this theory is that its proponents have attributed to the bird the ability to reason and deduce. But, as we shall see, the calculations used to determine one's position by a process of reasoning are extremely complex and an experienced (human) navigator, equipped with sextant, chronometer and astronomical tables, requires at least twenty minutes to find the answer. The experiments already quoted show that birds certainly do not proceed by a long series of mental calculations but somehow envisage the problem and reach an immediate and intuitive grasp of the answer.

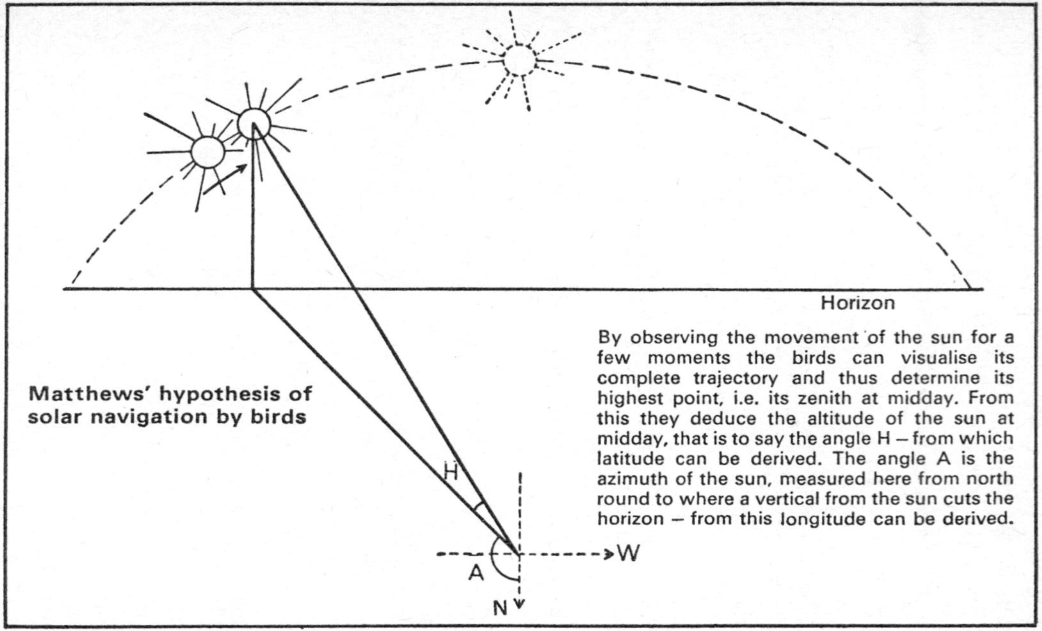
Matthews' hypothesis of solar navigation by birds
Horizon
H
A
N
W
By observing the movement of the sun for a few moments the birds can visualise its complete trajectory and thus determine its highest point, i.e. its zenith at midday. From this they deduce the altitude of the sun at midday, that is to say the angle H – from which latitude can be derived. The angle A is the azimuth of the sun, measured here from north round to where a vertical from the sun cuts the horizon – from this longitude can be derived.

Fig. 34

Let us look a little more closely at this problem of fixing one's position. A point is completely defined on the surface of the globe if its exact latitude and longitude are known. The movements of the sun are obviously not the same in the two hemispheres, but in the northern hemisphere between the North Pole and latitude 23° 54′ the sun not only appears to move from east to west but *its path is always to the south of us*; the reverse, of course, occurs in the southern hemisphere. On the Equator, the sun deviates to the north in 'winter' and to the south in 'summer' (the terms are purely conventional). Within the tropics the sun has a dominant deviation either to the north or to the south. Navigators determine latitude by measuring the height of the sun, i.e. the angle marked H in our figure. Providing the date is known, the angle H will give the latitude if the observation is made at midday. At any other hour of the day one must make two successive observations with an interval between in order to work out the latitude.

Knowledge of the angle A, the azimuth in our diagram, enables one to determine the longitude. From the observation one can deduce the local time and by comparison with the time at Greenwich one can convert the difference in time to a difference of degrees.

We can now take the case of a bird that wants by some means or other to keep a constant course during its journey. The rate of variation of the azimuth varies with the time of day and the curve representing this rate is modified throughout the year, so that the bird must have to compensate for the movements of the sun in order to hold a steady course. It would lose its way if it merely made a regular compensation of its orientation in respect of the sun (thus it is only at the poles during the equinoxes that the bird can navigate by modifying its relation to the sun by exactly 15° an hour). One can soon see that such navigational problems would be insoluble for birds if they were to be worked out by a process of reasoning.

While the theory of astronomical navigation in birds has been very fully discussed and argued by various workers, it is by no means resolved – far from it. The experiments of Dr Hans Fromme, for example, completely upset whatever intellectual complacency has been allowed to exist in the consideration of this much vexed problem. Fromme worked with robins, which we have already mentioned as being nocturnal migrants that fly to the south-west in autumn. But Fromme's robins surprised everyone by scorning the stars but holding to their south-west orientation even when the sky was totally covered. Better still, when they were transferred to a completely light-proof container they were still capable of finding the south-west. The method used, incidentally, was analogous to that used in the case of the warblers, the position of the bird being noted since robins also take up a flying position before actually taking off. But if the robins were then transferred to a container with six-inch concrete walls, they showed some hesitation, and they were often completely disorientated if they were placed in a container of steel. The robins appear, therefore, to be able to sense some mysterious radiation which is stopped by a layer of steel. If a small opening was made the birds were able to orientate a little better, presumably having perceived 'something'. When the steel casing was removed the robins, although still in darkness, were able to find the south-west alignment once again.

Science is at present baffled by this elusive radiation that the birds can sense but we are unable to identify. All known tests of radiation have been in vain. The influence of the terrestrial electro-magnetic field was quickly eliminated since neither a pile of iron nor an electro-magnet strong enough to deflect the earth's field by 90° had any effect on the birds. It has also been suggested that the birds are responding to cosmic rays, short wave radiations that come from stars and especially from many radio sources in the Milky Way. It is of interest that if one upsets the birds' internal clock (by progressive alterations in day and night length until the bird experiences 'days' of eight hours, eighteen hours, etc.) one does not at the same time upset its ability to orientate.

Another worker, Professor Helmut Adler, has published a critical study on the astronomical theory of bird orientation. He showed that an astronomical point, envisaged by a process of reasoning and deduction, would necessitate in the bird a physio-logical acuity of the senses, a concept difficult to comprehend. Adler believed that the internal clock in starlings is not as precise as people think, for an error of as little as five minutes in time can be transformed into a miscalculation of fifty miles or more in space. This strongly reinforces our own view. While such facts do not demolish the astronomical theory of bird orientation, they seem to confirm that, if the bird uses the deductive thought processes that man does, the precision of its calculations would be insufficient even to account for 'landing'. Just as the sight of an object imprints itself directly and immediately on our mind, so the perception of the direction that they take and the place where they are do not seem to entail a reasoning process in the brain of the bird but to take the form of an instantaneous and instinctive appraisal.

Fishes

Most fishes carry out migrations, but only some species are migrants. That sounds contradictory, but it all depends on the meaning attributed to the word migration. The second statement is true if one considers migration synonymous with long-distance movements round the world. Here, however, we refer to the definition given at the beginning of the book in which we have termed migration to be a cyclic and ordered displacement linked with external events (seasonal and cosmic rhythms) and internal events (reproductory and other cycles). In this case the first statement can be taken as true.

Take, for example, a population of carp in a lake. The carp make a true migration, for once a year the exigencies of the breeding season bring the fish together in a small part of the lake, always the same part, where they evidently find the conditions suitable for spawning (temperature 20°C amd so on). Breeding completed, the carp then disperse throughout the lake. This is a true act of migration, for it corresponds with the natural seasonal cycles. The migration is no less authentic than that of the tuna which congregate in the Mediterranean to breed and thereafter disperse as far as Norway to feed.

In the open ocean, boundaries are less obviously defined than on land, but certain species nonetheless confine themselves to quite strict territories which they travel through quite regularly. In the case of short-distance journeys, however, it is often difficult to decide what environmental factors are influencing the fishes since the factors involved are very variable. It is difficult, therefore, to distinguish between mere random wandering and true migration. If one tries at all costs to make some kind of division between the two and to label them as strictly migratory or strictly nomadic one can easily make a false interpretation of them.

An excellent example of this is provided by the Mackerel (*Scomber scomber*) if we consider the populations off the north coast of France. As a general rule, the mackerel avoid cold waters. In spring and in summer the waters of the North Sea begin to warm up and the mackerel come from the Channel and the open sea to spawn not far from the coasts. After spawning they come even closer to the shore, and one can catch them there in July to September in shallow bays and in the surface waters of the Channel and the North Sea. When winter comes the surface and coastal waters are considerably cooler, and the mackerel then go out to sea and into deeper waters where they find more constant temperatures.

To this one can add the fact that the most northerly regions are the first to begin to cool and the mackerel from the north begin to move before those of the south. This gives the impression that the mackerel are moving from the north to the south, whereas in fact they are merely leaving the shallow and coastal waters progressively from north to south. In spring the reverse occurs. This case is mentioned to show that one must interpret the facts properly and not jump to conclusions too hastily, for the mackerel appear to have migrated from north to south. Differing a little in detail, the movements of sardines and other small fishes have some relation to those of the mackerel, and will not be dealt with here. Instead, let us look at that most splendid of migratory fishes, the Salmon (*Salmo salar*).

We shall see how Dr Arthur D. Haslar managed to solve many of the mysteries surrounding the two phases in the life history of salmon, the first in rivers and the second in the sea. First, however, we should retrace the stages in the life of the species of salmon.

There are seven species of salmon. The Atlantic Salmon (*Salmo salar*) is the fish of European coasts and rivers. Salmon of the genus *Onchorhynchus* are found along the Pacific coasts of America and Asia and it was with such Pacific salmon that the most inportant results have been obtained. These fishes return in November and December from the sea to their rivers for spawning.

The spawning grounds that they seek out are small fast-running streams with gravel bottoms. Here the female digs a hollow trench or redd about two feet long, into which she deposits the eggs. The male, who is swimming close by, discharges his milt over the eggs and fertilization takes place. The eggs, which are translucent, round, about the size of a small pea, remain in the redd until the spring. The larvae that hatch out are minute and transparent amd the belly is swollen into a large vitelline sac containing the remains of the yolk. During this larval stage the fish is nourished by the yolk until it is finally resorbed and the young fry can then forage for small insects and crustaceans, for salmon are carnivorous. The fry increase in length and weight and the body becomes coloured. By the end of the summer we have a fingerling of two to three inches, amazingly quick as it flashes through the water. During the years spent in the fresh water the small salmon are called parr and have characteristic 'parr marks', dark blotches on the flanks. At the end of two years the fish is about six inches long. Only towards the end of the third or fourth year does the salmon, now called a smolt, descend to the sea. There are, however, many intermediates between the Pink Salmon (*Onchorhynchus gorbuscha*) of the American Pacific, which descends to the sea almost immediately after hatching, and certain individuals of the Atlantic salmon that may stay in the river until the seventh year.

Having descended the river and paused for a while in the estuary, the smolt then makes for the open sea. It now has a silvery underside and dark blue upper parts, the typical 'countershading' that counteracts the normal shadows and renders the fish less visible to predators. It would appear to be difficult to follow the movements of salmon once they are in the ocean, but such a study is possible with the use of various marking techniques. One of the most effective of these is the use of a small plastic tag which is tied by silver wire through the muscles of the back, the fishes being tagged as smolts during their descent to the sea. Such studies have been carried out on the salmon of the Pacific, the fish being recaptured by long lines strung with thousands of hooks. The map (Fig. 35) gives some indication of the scale of the migrations, although one of the most extraordinary cases is not shown. This was a pink salmon which was marked in the Gulf of Alaska and recaptured not far from Korea, a journey of 3,500 miles.

A study of a salmon population tagged in the Fraser River, on the Pacific coast of North America, showed that once in the sea

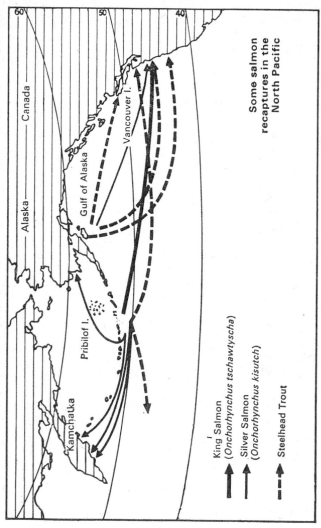

Some salmon recaptures in the North Pacific

King Salmon (*Onchorhynchus tschawtyscha*)

Silver Salmon (*Onchorhynchus kisutch*)

Steelhead Trout

Fig. 35

the salmon disperse in all directions, from the Gulf of Alaska as far as the Aleutian Islands. Methods of identifying populations of salmon are now very sophisticated so that one can tell which river a fish has come from (even if the fish is not tagged) by using physiological and immunological tests. The populations of salmon rarely mix, leading to a genetic isolation and thus the appearance of features that are unique to a particular population.

The salmon live in the oceanic waters for one, two, three or four years before returning to the stream of their birth, and it is here that the miracle occurs. For the salmon now manage to find the mouth of the very river that they descended as smolts up to four years previously, and this after considerable wanderings in the open sea in the intervening period. This is a staggering undertaking when one thinks of an indented coastline with numerous rivers and streams discharging along its length.

Dr Hasler has done more perhaps than anyone else to solve the mysteries of salmon migrations. He carried out a series of excellent experiments that demonstrated the ability of salmon to navigate by the sun.

It appears that fishes, like birds, ants, bees and crustaceans, are able to navigate using the solar compass. The phenomenon is thus fairly widespread in the animal kingdom. The first experiments were carried out on Lake Mendota fishes. Crossing such a lake presents a problem similar to that of navigating the open sea. It happens that the White Bass (*Roccus chrysops*) of Lake Mendota (Wisconsin, USA) spawns in certain places in the lake and not in streams. If one catches a fish and tags it and releases it at the farthest possible point from its spawning area, the fish will return with remarkable precision. One can even follow its track by attaching a marker to it at the end of a long nylon thread. The results of these experiments showed that, although solar navigation played the prime role in guiding the fish, other factors were used to help confirm it. The general principles of solar navigation are described in the section on bird orientation (see page 138). As explained there, the calculation of one's position using sextant, chronometer and astronomical tables is even now a lengthy process, taking perhaps half an hour. But the difficulties are enormously added to by the fact that the fish is in a liquid medium. This means that the fish must compensate for the fact that the rays of the sun do not reach it normally but are bent or refracted at the surface of the water. The relationship between the angle of incidence (i.e. the angle that the sun's rays make with the surface) and the angle actually observed by the fish is not a simple linear one. Double the apparent height of the sun and the angle seen by the fish is not automatically doubled but varies according to the equation proposed by René Descartes (sin i = sin r). Thus the sun's apparent movement is not 180° but appears to be only about 97° to the fish.

The experimental methods used to establish that fishes navigate by means of the sun are, of course, different from those used for birds, but the principle is the same. A fish is placed in a glass container that restricts its movements and this is in turn placed in a large tank which can be orientated or moved as required. There is a device that releases the fish from its container automatically and it is then free to swim in whatever direction it wants. Its subsequent movements are watched by periscope, for the presence of an observer might distract the fish and influence the results. This kind of apparatus was installed at the edge of Lake Mendota and the results, as in the case of those derived by Kramer on starlings, are highly significant. They proved that the fishes were using the system of the solar compass. The fish were found able to compensate exactly for the variations in azimuth of the sun in order to maintain a constant direction, and also capable of appreciating the height of the sun and thus knowing their latitudes. Thus, the fishes were equipped to select a course and to hold it throughout their journey.

It is, of course, quite legitimate to marvel at such faculties, but one must at the same time try to see them for what they are. What we feel to be marvellous and what in our sphere of reasoned and deductive thought appears comparable to the ability of a genius, is for the fish an unreasoned and immediate comprehension of its position and the direction to take in order to arrive at its destination. This is a kind of intuitive knowledge, difficult for us to understand who lack it. I would, however, draw the reader's attention to the case of those incredible human calculators who are capable of arithmetical feats whose speed rivals that of electronic calculators. Their ability is not merely a heightened version of the normal ability to add and subtract but seems to stem from a mental process as yet unexplored by science. What is interesting about such people is that, while they are making the

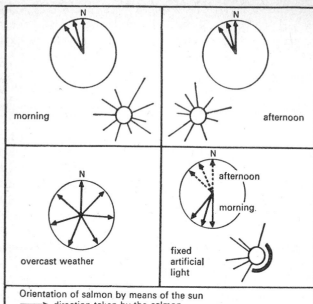

Fig. 36

calculation, they have no impression of doing more than exercising a perfectly normal faculty.

Returning to the Pacific salmon, there is another surprise in store. On reaching the mouth of their natal river, they then proceed to travel upstream and arrive in the very stream in which they themselves were spawned. This is a quite remarkable feat and hard to believe, but tagging experiments have confirmed it again and again. To appreciate such a feat one must imagine the often enormous hydrographic area covered by a particular river system. At the perimeter are the sources of the mountain streams, often hundreds of miles from the mouth of the river, and the whole river system forms a most complex pattern of branches. The salmon, nevertheless, ascends without hesitation. It is faced with a kind of maze and at each fork it has a choice between the correct channel and a wrong turning, but on each occasion it makes the right choice. If on rare occasions it makes a mistake, then it soon seems to realize this and retraces its steps to take the right turning.

For a long time the experts were baffled by this phenomenon. There was no dearth of theories to account for it, but none was satisfactory. It was suspected that the salmon were sensing something in the water, some physico-chemical stimulus that would guide them to their natal stream, but no one was able to find what that factor was. The prevailing opinion was that salmon require water with a high oxygen content for their breeding and that it was the oxygen gradient that they were following. This could not be true, however. A high oxygen concentration does indeed occur in the streams where they breed. But why, then, should salmon ascend barrages and waterfalls, often at great cost to themselves, when it is precisely below these that oxygen values

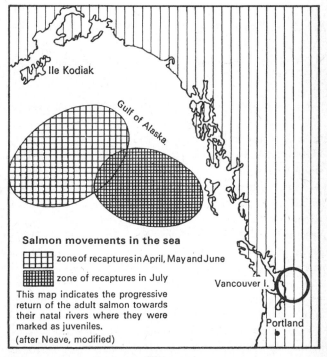

Salmon movements in the sea

▦ zone of recaptures in April, May and June

▦ zone of recaptures in July

This map indicates the progressive return of the adult salmon towards their natal rivers where they were marked as juveniles.

(after Neave, modified)

Fig. 37

are very high? Again, a factor as simple and widespread as that would not account for the way in which the salmon are able to find their way to a particular stream through such a labyrinth of channels.

The hypothesis which fits the observed data, but for a time remained merely an idea, was a very simple one. It was supposed that the salmon were able to find their way solely by means of an odour in the water which is unique to their natal stream. From the mouth of the river they recognize this characteristic odour and follow it to its source. This is rather similar to our recognition of houses in which we have lived by their characteristic smell, which we can recognize quite easily. For the salmon, the stream where it is spawned has its own odour as a result of particular local conditions such as the vegetation, geology, type of bottom, etc. There are small differences between each little stream, the bottom being slightly different, as also the plant communities along the banks and those in the water, and all these combine to give the water a special odour which the salmon are able to recognize after a number of years' absence (up to six years).

It remained, however, to test this rather daring hypothesis, for it presupposed that salmon must have an extraordinarily well-developed sense of smell. Again, it was Arthur Hasler, author of the hypothesis, who verified it experimentally.

First of all, this hypothesis is not incompatible with what can be deduced from the anatomy of fishes, for a large part of the brain is of cortical tissue that receives amongst others information from the olfactory nerves. It was then necessary to demonstrate the sensitivity of the olfactory organs. For this, salmon were trained to recognize a particular odour. The salmon were placed in a tank and trained to choose between two different odours, which we can label A and B. These were liberated in the water one at each end of the tank. The training was then carried out in the following way. If the fish swam towards odour A it received food, but if it swam towards odour B it was 'punished' with a slight electric shock. No other points of reference were possible in the tank (uniform walls, tank changed often, positions of A and B reversed, etc.). After some months of training it was found that salmon placed in the tank would swim without hesitation towards odour A. In nature it is probably from the egg stage that the odour of their particular stream is impressed on their memory. The acuity of the olfactory sense in salmon is quite amazing. Trained since birth to recognize a particular odour, they were capable of recognizing it in a concentration of as little as 3×10^{-18}. This represents the kind of dilution one would get from stirring a cubic centimetre of gin (about a thimbleful) into a body of water sixty times the size of Lake Constance! Two or three molecules of the odour are sufficient for the fish to be able to appreciate it.

The second part of the experiment consisted in proving that it was indeed the odour of their natal stream that attracted and led the salmon during the return to the stream. To this end, salmon were caught and marked some three miles above the branching of two channels of the river. One knows, therefore, which branch they will take. They were then released below the branch of the river and when they were recaptured again it was invariably in the channel they had chosen the first time. But, to prove the point, if the olfactory system was in some way put out of action, the salmon then returned about equally to the channel they chose the first time and the other channel, proving that it was their powers of smell that guided them.

The ascent to the spawning grounds is also a tremendous physical performance, and the salmon is commonly shown making its way upstream and leaping waterfalls with incredible power. 'It is surely one of the most perfect of creatures to watch; clothed in silver of the greatest purity, the curve of its body speaks of a subtle grace, a concentration of energy and a perfect adaptation to intense activity,' wrote F. M. Duncan.

Barrages and turbines that have been newly constructed are often a catastrophe for salmon, for the fishes will try to jump higher and higher to get past the obstacle, often exhausting themselves and dying. In some cases, usually at the insistence of a department concerned with fishery protection, the contractors are required to install a fish ladder, an arrangement of pools flowing one into the next so that the leap between successive pools is not too great.

After this display of wild energy and superb elegance, the salmon arrive at the spawning grounds. They find their ancestral home, where they were born and where they will die. Here they reproduce. Since their entry into the river the salmon have not fed, all their efforts being devoted to ascending the river and their entire metabolism being used to prepare for reproduction. The gonads have enlarged considerably and now occupy a large part of the body cavity, at the expense of the digestive organs. Thus the ovaries of the female can represent as much as a fifth of the weight of the fish in a salmon of 20 lb. After the female has deposited the eggs and the male fertilized them, the salmon have nothing left but to die. Generally they do not return to the sea but remain in the river, enjoying a well-deserved rest. They lose their fine colours and their robe of silver scales and become lustreless.

While the salmon of the Pacific generally do not survive after spawning, those of the Atlantic live on and, as kelts, pass down the river to spend a further period in the sea. The Atlantic salmon, which we have not dealt with here, is quite as remarkable a migrant as its relatives in the Pacific. Experiments on the homing powers of these fishes have shown that it can regain its natal river even though it may have wandered some 1,500 miles away from it. One individual is even recorded as having travelled 2,500 miles between the south-west of Sweden and the west coast of Greenland. The Atlantic salmon can reach far to the north and even enter polar waters.

It can be noted that salmon, with their highly developed olfactory powers, are also very sensitive to chemical pesticides which have been indiscriminately used. For example, DDT (to cite just one and perhaps one of the least noxious) has decimated numerous populations of salmon, causing blindness, sterility and finally death. Here is a case of the idiocy of farm and forest managements, aided by unthinking chemists and the wickedness of a system that allows vested financial interests to promote new forms of destruction solely for the profit of the shareholders – a system that has no regard for nature. This shows the poverty of thought behind campaigns designed to do battle with insects. To believe that one can selectively kill certain insects while sparing others is folly. Water transports the insecticides to another place, and one animal species feeds upon another. The insecticides accumulate in the fatty tissues of living organisms, become concentrated there, and build up to toxic levels which kill off the predators of those species.

Man himself is far from being safe in the future from all the consequences of his present actions in regard to other animals (malformation, sterility, toxic levels of pesticides, etc.).

Trout rivers and lakes do not always provide all the spawning requirements of these fishes (high oxygen levels, freedom from predators, and so on), and in springtime the trout ascend affluent streams to breed. There are, however, many trout species and subspecies that are sedentary. In this context, one should not really speak of 'trout' and 'salmon' as if there were a hard and fast distinction between the two. Thus the Brown Trout (*Salmo trutta*) and the Salmon (*Salmo salar*) belong to the same genus, as also the Cutthroat Trout (*Salmo clarki*) and the Steelhead Trout (*Salmo gairdneri* – of which one subspecies that remains permanently in fresh water is known as the Rainbow Trout). All these species are more closely related to one another than they are to the Pacific salmons (genus *Onchorhynchus*) or the Brook Trout (*Baione fontinalis*).

Although the migrations of sturgeons are as well known generally as those of the salmon, the details are not so clear. The fry hatch in large rivers in spring and at the end of the summer make their way to the sea. Their movements in the sea are not precisely known. The adults are splendid fish, six to nine feet long but reaching as much as twenty feet in the case of the Russian sturgeon or Beluga (*Acipenser huso*) of the Black and Caspian seas. The female is escorted by a number of males which are smaller than the female. Ascending the river, the female lays its eggs – the famous caviar – in a depression which it digs in the river bottom. Altogether, sixteen species of sturgeon are known, occurring in North America and Canada, Europe and Asia. Sturgeons may be found some distance from the sea but as a general rule they do not penetrate far inland. Some, such as the Lake Sturgeon (*Acipenser fulvescens*), spend their entire lives in lakes and rivers and do not migrate to the sea.

As we have seen, salmon are bred in fresh water and then migrate to the sea and return to fresh water again only to breed. In contrast to this anadromous pattern of life, the European Eel (*Anguila anguila*) is catadromous and does the reverse, breeding in the sea and then migrating to rivers for its major period of growth. Eels and salmon, in their different ways, are perhaps the most celebrated of migrant fishes.

The European eel breeds in the Sargasso Sea between 22° and 30° N and 48° and 65° W. This was the sensational discovery made by Professor Johann Schmidt as a result of a most painstaking and detailed study of the biology of eels. Man had noticed and pondered on the sudden appearance of young eels in rivers but their spawning grounds had hitherto remained a mystery until Schmidt's classic studies in the 1930s. Schmidt reached his conclusion by compiling statistics of the length of eel larvae or leptocephali from different points and at different times in the Atlantic. Plotting these on maps, he was able to join together points which represented places where eels of the same length had been recorded. The result was a series of zones marking leptocephali of decreasing sizes and at the centre, where the smallest larvae were found, one would expect to find the breeding area. This was in the Sargasso Sea.

The eels spawn at some depth at the beginning of spring but the young larvae do not rise to the surface until the beginning of summer. The smallest of them measure only five millimetres and at this stage they are feeding on the plankton around them. These leptocephali are quite unlike the adults, for they are thin, leaf-like and transparent. Their movements are determined by the ocean currents. If they drift too far north they reach the Labrador Current and are killed by the cold water. If they pass into the Gulf Stream, however, they can slowly drift across to Europe (see Fig. 38). One can even plot the course of these currents by the presence and distribution of the leptocephali in the Atlantic. Keeping about 60–150 feet below the surface, they feed voraciously on small planktonic animals and reach three-quarters of an inch in the first year and about three inches by the end of the third summer, by which time they have drifted to the shores of Europe. Once there, they make a quite remarkable transformation. From the leaf-like leptocephalus, they change into a cylindrical animal much more nearly like the adult eel. At this stage they are referred to as elvers and resemble tiny pearly-white replicas of their parents. This is a time of profound physiological and morphological change and the animals do not eat until the transformation is complete. Soon they leave the deeper water, make for the shore and enter the estuaries of rivers. They are now subject to two definite tropisms or reactions to specific stimuli. The first is positive rheotropism, which is a response to water currents such that they will swim against any current that they encounter (quite different from their behaviour in the sea). Even the slightest current is enough to stimulate the

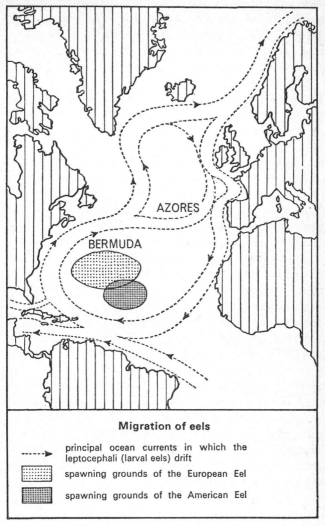

Migration of eels

- - - → principal ocean currents in which the leptocephali (larval eels) drift

spawning grounds of the European Eel

spawning grounds of the American Eel

Fig. 38

elvers to fight against it, and in this way they will sooner or later reach the source of the river. The other factor which affects them is chemotropism, and in this case the elvers react positively towards fresh water and avoid salty water. The changeover from salt to fresh water presents many physiological problems, as it does also in salmon, and the way in which such small animals can adapt themselves to this change is not yet fully understood.

The elvers make their way far inland, wriggling up waterfalls and little channels and even crossing fields on rainy nights. The female eels grow to a large size (up to about four and a half feet in exceptional cases) but the males are smaller and rarely reach more than eighteen inches in length. It can be mentioned here that sex is determined less by genetic factors than by conditions under which the animal lives so that the hermaphroditic elvers may become either males or females depending on circumstances.

For many years the eels live a quiet and sedentary life in the place that they have eventually chosen. The years go by and, preying on crabs, small fishes and other animals in the rivers, the eels grow into extremely powerful fishes (as anyone knows who has handled them alive). It is the males who first give up this tranquil life in order to make the long journey back to the Sargasso Sea. They are at this time between eight and fourteen years old and have thus spent between five and eleven years in fresh water. Those that remain die. The larger and even more powerful females, which have been known to live for over twenty years, only begin the long journey when they are ten to fourteen years old. It seems that the return to salt water is a very urgent need for eels for some physiological reason. Not only do they change outwardly, becoming silvery and being known as silver eels, but the whole hormonal system undergoes a tremendous change (the gonads in particular develop considerably). In autumn the eels depart and the migrants, abandoning their lake or quiet reach of river, make for the mainstream or move across fields to streams while the fields are wet with dew or rain. The eels have a tremendous ability to resist desiccation, partly as a result of the copious mucus exuded from the skin, and they can remain out of water for hours or even days. During such overland journeys, the eels find some channel or damp hole where they can rest and avoid complete desiccation. They breathe by retaining a certain amount of water round the gills and thus preventing the fine gill filaments from drying out altogether. Once they have reached a river they descend to the sea and must make the physiological adjustment to salt water. Freshwater fishes tend to absorb water because the concentration of salts in their body fluids is higher than the surrounding medium. Saltwater fishes, on the other hand, tend to lose water since the salt concentration outside their bodies is higher; they thus tend to drink copiously. The salt/water balance is regulated by the kidneys and the physiological problems raised by anadromous and catadromous fishes have yet to be solved satisfactorily.

Once in the sea, the eels are presumed to make their way back to the Sargasso Sea to breed. This is presumed because eels have not yet been caught in the open sea. This fact led Dr Denys Tucker to propose an alternative theory. He suggested that in fact the European eels never actually return to the breeding grounds. How then is the species perpetuated? Tucker's ingenious answer was that a certain proportion of the American Eels (*Anguila rostrata*), which also breed in the Sargasso Sea, are drifted to the American coasts, but that another part of the population are those that pass into the Gulf Stream and are drifted to Europe. He explained the difference between the two supposed species of eels, which is mainly in the number of vertebrae (slightly less in the American eel), as being due to the difference in temperature in which the eggs and larvae develop. It is, in fact, well known that members of the same species can have rather fewer vertebrae if hatched and grown under warmer conditions (salmon, herring, etc.). Until a migrating European eel is found in the Atlantic Tucker's theory must remain, if not proved, then at least not discredited.

The migration of the leptocephali on their journey of three years and 3,000 miles is still a remarkable fact, even if there is some doubt that the adults do the same. One rather audacious theory, but one which it is hard to see how to prove, suggests that the Sargasso Sea was once an inland sea on the 'lost' continent of Atlantis, into which flowed all the rivers of the continent. Thus the eels, still with some inborn behaviour pattern, attempt to return to their ancestral breeding ground. Another theory supposes that the American and the European continents have slowly drifted apart and that what was once a relatively short journey for the European eels has now stretched to 3,000 miles.

The Grey Mullets (species of *Mugil*, etc) and the Sea Basses (species of *Dicentrarchus*, etc) belong to the same group as the eels (i.e. catadromous). They breed in salt water but spend the rest of the year in coastal waters, estuaries and even in fresh water (especially mullets).

The tunas are rather exceptional fish; they are not strictly cold-blooded vertebrates like other species, for the temperature of their internal organs rises to as much as 8°C above that of the surrounding water. To maintain this temperature difference, the tunas have a more highly developed circulatory system than other fishes. The tunas belong to the family of mackerel-like fishes (*Scombridae*) and there are six species of great tunas in the world: Albacore (*Thunnus alalunga*), Yellowfin (*T. albacares*), Blackfin (*T. atlanticus*), Bigeye (*T. obesus*), Bluefin (*T. thynnus*) and Long-tail (*T. tonggol*). Two of these are of particular interest to us here, the bluefin, which is the main species that is tinned and is usually referred to merely as the tunny, and the albacore.

The albacore, which can be recognized by its long sickle-shaped pectoral fins that reach back to the beginning of the second dorsal fin, is a species that breeds in the tropics, both in the Atlantic and the Pacific.

Let us deal with the Atlantic populations first. Having spawned, the adults and those one-year-old fishes that have not yet left tropical waters depart in February for the north. It has been found that this northward movement is by no means haphazard. E. Le Danois showed that the migration of albacores is strictly related to water temperatures, and he discovered the following rule. 'In the course of its journey along the coasts of Europe, the Albacore does not leave water whose temperature at a depth of 150 feet is equal to or above 14°C.' This very clearly defines the habits of the albacore.

Thus, during the spring, the albacores move northwards as the 14° isotherm progresses north until, by the end of May, they are in the Bay of Biscay, where tunny from both Spanish and Breton waters await them. By September the albacores reach their northern limit, which is to the south-west of Iceland. Surface temperatures drop a few degrees during the night in these waters and the tunas accordingly dive down to depths where the temperatures remain constant. For this reason, the tunas are not fished for at night. In the autumn, when the entire water body becomes too cold, the albacores head southwards but all the time swim further from the surface until by winter they have reached depths of 900–1,200 feet in tropical waters. The indigenous fishermen of the Azores and the Canary Islands catch them at this depth with lines.

In the Pacific there are comparable migrations, and the migratory movements in the northern part of the Pacific have been particularly studied. Thus, there is good reason to think that it is the same population that occupies the whole of the North Pacific, from American shores to Japan. In the middle of this huge area the most complex migration patterns occur and these are shown in Fig. 39. In spring, the six-year-old tunas move southwards to breed in tropical waters. After the larval stage the young of the year move towards the temperate waters of North America, but they escape the fishery until they are two years old. From that age they take part in the regular east-west migrations. From small begininngs, these migrations become more and more extensive until they reach right up to the Japanese coasts. Until the fishes are three years old, those that have been marked in America go westwards in August and September, but for the most part return to American waters in the following summer. In the fourth year, however, the same migration is followed but this time it is only a minority that return to the East Pacific, the majority then entering the Japanese circuit. At six years old the albacores migrate southwards to breed in tropical waters.

The north-south movements of these fishes are controlled by water temperatures, as in the Atlantic populations.

The tunny or bluefin is a much more imposing fish. It weighs from 300 to over 1,000 lb. and may reach a length of twelve feet (the record rod-caught fish was 977 lb. in weight). It is a superb animal with a dark blue back set off by silver flanks flecked with gold or opalescent spots on the fins. It seems to fear neither sharks nor whales and is a born migrant. During the breeding season it congregates in very well defined areas and at that time ceases to feed. With spawning accomplished, the bluefin is no longer a breeding tuna and becomes a wandering tuna, swimming off in search of food and travelling perhaps thousands of miles.

The breeding tunas arrive on their spawning grounds in April and May. The best known of these areas is that which is bounded by Tunisia, Sicily, Sardinia and the Mediterranean coast of

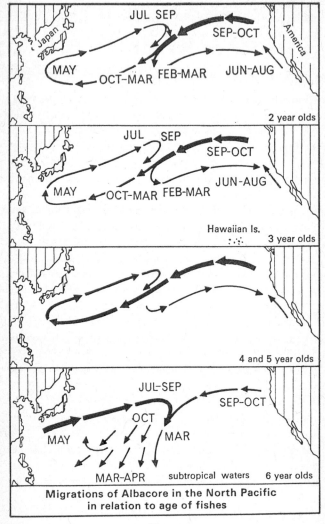

Migrations of Albacore in the North Pacific
in relation to age of fishes

Fig. 39

Spain. Similar areas exist in the Black Sea, the Atlantic just outside the Straits of Gibraltar, in the region of Bermuda and in parts of the Pacific.

To enter the Mediterranean, the bluefins must pass through the Straits of Gibraltar. After Lozano Cabo, the fishes show a markedly negative rheotropism, swimming wherever possible with the prevailing current. In the Straits of Gibraltar, super-imposed on the main tidal current, is another current flowing in from the Atlantic to compensate for the water lost by evaporation in the Mediterranean. This current is strongest in the middle of the straits and it always dominates the tidal currents. Thus the tunas migrate in the middle of the straits. Within their spawning area the bluefins follow absolutely set routes which have been known to man since antiquity. These routes often pass close to the shore and the fishermen set up a system of nets known as *madragues* at the same places and at the same times each year to catch the tunas. The catches are indeed enormous and the for-tunes of many Mediterranean ports have been founded on the tuna fishery.

After spawning, most of the bluefins move out into the Atlantic, although it is believed that there must also be a sedent-ary population that does not leave the Mediterranean. Having passed Gibraltar, often in huge shoals of up to 10,000 fishes, the bluefins pass up the coast of Spain and France and into the Channel (by about June). This was once their northern limit, but recently (since 1921 to be exact) the bluefins have been passing northwards as far as Norway, where they are now fished in some numbers from July to October. To reach so far north, the blue-fins must cover not less than 3,000 miles, and this in the space of scarcely a month.

During this time, the bluefins of the Atlantic coast of North America are also moving northwards, but their Pacific coast relatives that move along the Californian coast differ in that they move southwards in spring and northwards in autumn.

There still remains a great deal to learn about the movements of the bluefin in the Atlantic. For example, it is known that some of them actually cross the Atlantic, as has been shown by recap-tures of fishes marked in the Bahamas and subsequently caught in the Gulf of Gascoigne. It is not yet known, however, whether a major part of the population takes part in such migrations or

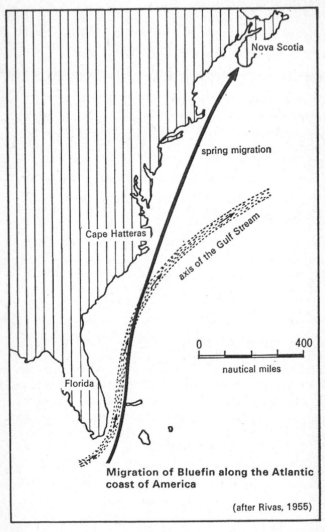

Migration of Bluefin along the Atlantic coast of America

(after Rivas, 1955)

Nova Scotia

spring migration

axis of the Gulf Stream

Cape Hatteras

Florida

0   400

nautical miles

Fig. 40

whether this merely represents a few more adventurous individuals.

Elsewhere in the world other populations of bluefins undertake regular migrations, travelling along Australian coasts, for example, some 1,200 miles.

Some of the smaller relatives of the great tunas also make migrations. The Oceanic Bonito (*Katsuwonus pelamis*) is a species that grows to about three feet in length and is a handsome fish with a remarkable turn of speed (about twenty-five miles an hour) which enables it to chase flying fishes, often leaping clear out of the water to do so. Bonito are also said to hunt in groups, encircling a shoal of fish. After spawning, the bonito of the Black Sea and the Sea of Marmora move out into the Atlantic and become widespread between the latitudes 55° North and 35° South. Like the albacore, the bonito is very sensitive to water temperatures and 20°C seems to be its optimum. A related

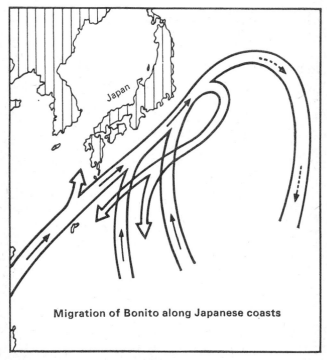

Migration of Bonito along Japanese coasts

Fig. 41

species, *Euthynnus alleteratus* (known as the *thonine* in France), avoids colder and less saline water and seems to be generally more susceptible to cold in the Indian Ocean since it is absent when temperatures fall below 22°C.

Another monarch of the seas is the Swordfish (*Xiphias gladius*) and this species must also be a great migrant. Unfortunately, not a great deal is known of its movements. It is thought to breed in tropical waters and then to spread all over the globe, including even the Arctic seas.

Amongst the numerous species of sharks in the oceans there must surely be some that rank as great migrants. But sharks are not fished to the same extent as other fishes and have not been subjected to intensive marking experiments, so that one can only generalize and say that many species breed in shallow waters and then must disperse into the open seas in search of their prey.

The Herring (*Clupea harengus*), the fishery of which has had a profound economic effect on the countries bordering the North Sea, has for a long time been considered a migratory species. This idea arose through observations on the movements of the fishing fleets and the areas where they set their drift nets. The fishery begins in summer in Norwegian waters and then moves progressively south and west until it reaches the Channel in December. In fact, the herring is widespread throughout the North Atlantic, especially over the continental shelves, and it seems to have a nomadic pattern of behaviour largely linked to the salinity of the water. The herring seem to prefer to keep to the boundary between the polar waters which are cold, dense and sluggish, and the tropical waters which are warmer, less dense and faster moving.

The movements which the fishermen believed the herring undertook result from another phenomenon. During the spawning period the herring, which had until then been widely dispersed, congregate in extremely large shoals. Since the time of spawning is not the same along the coasts, but moves progressively down the coasts, the herring catches give the impression that it is the same population that is moving southwards.

The Sardine (*Sardina pilchardus*) lays its eggs in water of 10°–17°C, depending on the time and place. The movements of sardines are very complex and difficult to unravel.

The European Anchovy (*Engraulis encrasicolus*) is more particular and requires a temperature of at least 17°C before it will lay its eggs and fertilize them. The areas where it will breed are limited to those which are shallow and warmer than the main ocean. In the North Sea, for example, it used to invade the Zuider Zee, entering the estuaries of rivers like the Elbe and the Scheldt. In the summer, the anchovy reaches as far north as Bergen but it is driven southwards by the cold as autumn approaches and it returns to the Atlantic. It is known from the coasts of North Africa and there is a smaller form that occurs off West African coasts and spreads southwards to meet the South African Anchovy (*Engraulis capensis*). The population off West Africa breeds there and therefore does not represent winter migrants from Europe.

The Cod (*Gadus aeglefinus*) is like the tunas in having a feeding area very much larger than its breeding area, which is always a strong indication of migratory behaviour. It migrates each spring to particular spawning areas, especially off the Lofoten and Faroe Islands, Iceland and Newfoundland.

The list of migratory fishes is a long one but, except for some of the more spectacular species, much less is known of migrations of fish than is the case with birds. The physical difficulties of observing fish movements are mainly responsible for this. As we have seen, observations based on fishery records can be most misleading. Methods of marking or 'tagging' individual fishes are much more reliable and now form standard practice in fishery investigations. There is no doubt that a great deal will be learnt in the next decade as a result of such work.

Insects

Leaving aside such spectacular phenomena as locust invasions, it is true to say that on the whole insect migrations are less well known than are migrations in other animal groups. That butterflies can be long-distance migrants comes as a surprise to many people, so widespread is the belief in the extreme brevity of their lives. It is commonly thought, for example, that after a period spent first as a caterpillar and then as a chrysalis, the perfect insect finally emerges into a world that it is destined to enjoy for only a few days. There are indeed some insects, and notably the Mayflies (*Ephemeroptera*), whose lives as adults are very short, the adult emerging and living only for a single day so that it can reproduce. But in most insects the life span can be measured in weeks, months or even years, so that they are quite able to travel considerable distances. The migrations of insects have certain curious features. In fact, while the life cycle of migratory species is not short (relatively speaking), it nevertheless often occupies less than a full year. The year is, however, the basic time interval in all migratory cycles, so that in an authentic migration, with a definite outward and return journey every year, it would not be (save for exceptions) the same individuals that make both journeys.

A question immediately springs to mind. How can the second generation then return to the same place from whence its parents came if the latter are not there to guide them? However, just as young storks have been found to follow the parental migration route even if isolated and released after the parents have flown (see page 129), so the same happens in insects, presumably also obeying an instinctive pattern of behaviour.

Our knowledge of the migratory behaviour in insects is derived from an eminent but rather small body of research workers (one can mention especially C. B. Williams, Sir Boris

Uvarov, Dr F. A. Urquhart). Although the group *Insecta* embraces an enormous number of species, more in fact than any other animal group, there have on the whole been fewer studies of the migratory behaviour of insects than of such groups as birds or fishes. Nevertheless, records of certain insect migrations date back to earliest times, and the best known concerns the locust invasions that are recorded in the Old Testament (Exodus 10, verses 12–15):

And the Lord said unto Moses, Stretch out thine hand over the land of Egypt for the locusts, that they may come upon the land of Egypt, and eat every herb of the land, even all that the hail hath left. And Moses stretched forth his rod over the land of Egypt, and the Lord brought an east wind upon the land all that day, and all that night; and when it was morning, the east wind brought the locusts. And the locusts went up over all the land of Egypt, and rested in all the coasts of Egypt: very grievous were they; before them there were no such locusts as they, neither after them shall be such. For they covered the face of the whole earth, so that the land was darkened; and they did eat every herb of the land, and all the fruit of the trees which the hail had left: and there remained not any green thing in the trees, or in the herbs of the field, through all the land of Egypt.

This was the eighth plague, but even this, we are told, did not soften Pharaoh's heart although the effects on the agricultural community must have been devastating. Nowadays there is an Anti-Locust Research Centre in London where all information on locusts, and particularly the latest reports on their movements, is collated so that effective action can be taken. Since the swarms may originate in one country but spread over many others, it is essential that control be organized on an international scale. It was largely due to the efforts of Dr Uvarov that such a centre came into being.

The migratory species of locust resemble grasshoppers and other members of the order *Orthoptera*, and are distinguished by their large hind legs and short, stumpy antennae. They have two pairs of wings, the anterior pair heavily stiffened with chitin and folding over the hind pair when not in use. The seven principal species of locust all show a migratory disposition, although in varying degrees. The greatest depredations are mainly the result of two species, the Desert Locust (*Schistocerca gregaria*) and the Migratory Locust (*Locusta migratoria*). In addition to these, there is the Moroccan Locust (*Dociostaurus maroccanus*) which occurs from the Canary Islands to the Caspian but does not make very long journeys, and in the southern part of Africa there is the Red Locust (*Nomadacris septemfasciata*) and the Brown Locust (*Locustana pardalina*). The latter spreads out from time to time from the interior southwards. The southern part of the American continent is the habitat of the South American Locust (*Schistocerca paranensis*), while the north of the continent has in former times had to contend with the depredations of the Rocky Mountain Locust (*Melanoplus spretis*). The latter has, however, disappeared, not having been seen since the turn of the century.

During the last century the Rocky Mountain locust carried out enormous invasions which covered almost all the green areas in temperate North America. Its migrations have left some curious traces. In Montana, for example, one finds dark lines on a glacier on Mount Cook at 11,000 feet, and on closer inspection these turn out to be the remains of millions of locusts caught in the ice. The explanation is simple. The locusts, like other aerial migrants, follow special routes through the mountains. These are based both on topography and the distribution of air currents, the locusts finding the routes which afford the least possible effort during the ascent (rather like the bison trails mentioned earlier). But cold and hunger overcome large numbers of migrants and their bodies drop on to the glaciers, where they are covered by the falls of snow in the winter. In this way a virtual sediment is built up year after year and this appears as a dark band on the mountain side. Using the radio-carbon (c14) dating technique – basically, an assessment of the amount of decay of this carbon isotope, since its rate of decay is known – attempts have been made to date the locust deposits. The results suggest that bodies of the locusts have remained intact in the glaciers for several centuries (this is, incidentally, a useful method to date the age of the glacier).

A word now about the life cycle of locusts. In the breeding season the females dig a hole in the sandy soil and for this purpose use their ovipositor, a slender prolongation at the hind end of the body which is thrust into the earth and then helps to guide the eggs into the hole (in bees and wasps the ovipositor has the function of a sting – which explains its absence in the male). The female lays about fifty to eighty eggs and these are enveloped in a frothy secretion which forms an egg pod and prevents them from drying up. Eggs may be laid many times in the season. The young emerge from the egg, not as caterpillars but as nymphs or hoppers, i.e. fairly exact replicas of the adults but lacking wings. Although they are thus incapable of flight, the aptly named hoppers are extremely active and can walk and jump in a very determined fashion.

The case of the migratory locust is a little complicated in some respects, since there are three breeding periods a year. Hatching takes place after the first spring rains and the hoppers, scarcely out of their eggs, straightway start moving northwards. They form small groups which continuously grow as more recruits join their ranks, until finally there is a huge living carpet that progresses slowly across the ground surmounting any obstacles lying in its path. They invade towns that lie on their route, covering the streets and entering houses. The way in which they cross water-courses is very spectacular. When they seem to have found the most favourable crossing, those in front simply throw themselves into the water and form a living bridge for their comrades. The locusts are not easy to drown, and an uninterrupted rolling and wriggling between those in the water and those that are walking over them ensures that the living bridge survives and stretches across to the other side. The surface of the water is thus covered by a scrambling, writhing mass and in fact many are drowned in the process, but this makes no appreciable difference to the multitudes that succeed in crossing. One can see that the locusts have no need to wait until they are adults in order to migrate and survive the hazards of a long journey. But when they do acquire wings their invasions are even more spectacular.

At first sight, these invasions seem to be of a different order to migrations in the strict sense since they appear to have no correspondence with the seasons. One should not, however, hold to such a strict definition and if evidence can be found to show that the cycle involved is a complex one stretching over several years, then these movements can be reckoned as true migrations.

The desert locust shows a quite classic migratory pattern of behaviour. In effect, its regular movements correspond to the two generations produced by this species every year. Dr J. S. Kennedy was able to show that the limit of the two areas of breeding was strictly determined by the position of the

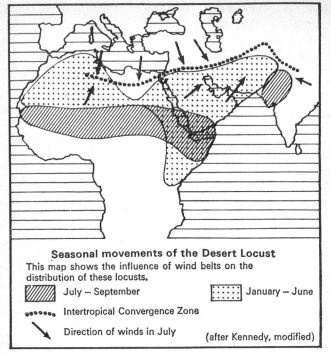

Seasonal movements of the Desert Locust
This map shows the influence of wind belts on the
distribution of these locusts.

▨ July – September   ▦ January – June

●●●●●● Intertropical Convergence Zone

↘ Direction of winds in July   (after Kennedy, modified)

Fig. 42

Intertropical Convergence Zone. This front, which moves with
the seasons, marks the boundary between two opposing wind belts
(see Fig. 42). In July, for example, the locusts do not penetrate
north of the front, for if they did they would encounter north
winds that would force them back. Most important in this
connexion is the fact that any sudden changes in the position of
the front immediately affect the distribution of the locusts. (It
is exactly this kind of information that is sent to the Anti-Locust
Research Centre so that prompt action can be taken.)

Europe has also been afflicted with locusts and there are
numerous accounts of locust invasions that have reached central
Europe, Hungary, Germany and even France during the Middle
Ages. Nowadays, locusts that are seen in France come mainly
from a large breeding area in Persia, but may also come from the
south-west of France. A few stragglers reach the British Isles but
they are unable to survive the cold and do not breed.

The breeding areas of the Middle East were responsible in olden times for many European invasions. The following is an early and very graphic account of one such invasion:

I saw the plague for many consecutive years, particularly 1645 and 1646. These creatures not only come in legions but in entire clouds, in clouds five or six leagues long [i.e. twelve and a half to fifteen modern miles], coming generally from Tartary. This happens during a dry spring. These vermin are blown by the east or south-east winds towards the Ukraine, where they do great damage, eating all kinds of crops and grasses so effectively that everywhere that they have passed in less than two hours they have cropped everything they find, causing great loss of lives . . . It is not easy to estimate the numbers since the air is full and loaded with them. I do not know how better to describe their flight than to compare it to snowflakes driven by the wind in stormy weather, and when they settle in order to feed, the fields are covered with them . . . When they rise they are carried by the wind and when they fly, although the sun is still shining, it is no lighter than in the cloudiest of weather.

At some point in the history of very many countries there have been invasions by locusts. The populations of locusts fluctuate widely, while the area inhabited by them can also vary in extent within quite a short space of time, mainly as a result of environmental factors. In turn, this affects the balance between the level of the population and the resources available to it. In the migratory locust, for example, there was a very considerable extension of its distribution around 1930 (Fig. 43), whereas there was a rapid decline of the desert locust.

The wave of migrating locusts may pass like a compact and fluid mass close to ground level, but one may see another layer of locusts about thirty feet above the first and moving in a different direction. The swarm, stirred by air currents such as rising thermals, can sometimes take on the appearance of gigantic spirals. Generally, a certain direction is held for many hours and then suddenly the swarm flies in quite another direction. The actual flying speed does not exceed that of a man walking fairly fast, but this may be augmented by the wind.

It is naturally difficult to estimate the size of a cloud of locusts. Nevertheless, a figure of no less than 50,000,000,000 has been advanced for certain particularly dense swarms.

What are the causes of these locust migrations? At first sight it does indeed seem that the exodus of locusts defies any rigid classification, for one is faced with the fact that periodically there appear huge swarms of migrants of which there is no sign at other times. However, some very careful observations by Uvarov showed that what had been taken for different species of sedentary grasshopper and migratory locust were in fact different *phases* in the life of the same species. What had seemed an insoluble problem was all at once clarified by Uvarov's 'phase theory', published in 1921.

Uvarov suggested that the insect passes through a series of phases during its life cycle whereby it changes from one form to another. In the case of the locust, this means a transition from a sedentary form to a migratory one, from a grasshopper to a locust.

During the solitary phase the locusts live independently from each other and during this period they are in biological equilibrium with their environment and with their natural enemies. The effect of their feeding on the vegetation is no different from that of other organisms and is in no sense dramatic. After a transitional phase, the locusts then enter the migratory phase and their behaviour is radically altered, for they now assemble in large numbers and display an intense migratory behaviour. Each phase may last for a longer or shorter period, the length of each being strictly dependent on the size of the population, or rather its density. This is, in fact, the kernel of the problem. When the density of a sedentary population reaches a critical level, then the individuals pass into the gregarious and migratory phase.

While locust invasions are catastrophic as far as the vegetation is concerned, they also are the finish for the locusts themselves, for a mass of insects as great as this cannot be in equilibrium with the resources available. The swarms do not always find adequate vegetation in the new areas but even when they do find favourable conditions the swarm often presses on regardless and ends up in some quite arid spot where there is definitely not enough food to go round. Thus the journey itself and the obstacles encountered on the way lead to a veritable hecatomb of locusts.

In this way, the migratory phenomenon seems to be a regulatory mechanism, a sort of counter-balance, which limits overpopulation and restores the normal balance between the insects and their food supply. But the search for food cannot entirely explain the exodus since the area where the nymphs hatch out usually provides sufficient food, whereas the end of the journey is often little suited to the needs of this huge band of migrants. Before undertaking their journey, however, the locusts congregate in certain very definite 'outbreak areas' which are distinct from those of the solitary form. The characteristics of such outbreak areas vary with the different species. The migratory locust chooses marshy areas and reed beds (e.g. flood plains of the middle Niger and in central Asia round the Caspian and Aral Seas), while other species choose semi-desert regions where there is merely a covering of grasses on which the nymphs can feed (e.g. the southern edges of the Sahara). The migratory activity which is manifested at these outbreak areas can be calmed by hordes of predators drawn to such a rich supply of prey. They reduce the numbers of locusts and thus reduce the population density, with the result that the principal stimulus to migration is removed. In some ways this touches on the problem of the psychology of crowding, as we saw earlier in the discussion on rat populations and lemmings (page 26).

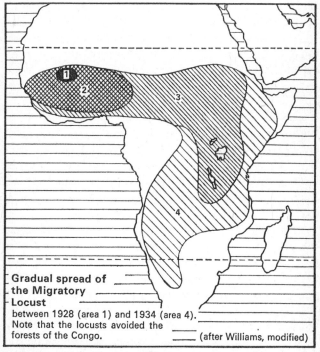

Gradual spread of the Migratory Locust between 1928 (area 1) and 1934 (area 4). Note that the locusts avoided the forests of the Congo.
(after Williams, modified)

Fig. 43

Much remains to be learned about locust migrations. It seems certain that changes in the balance of hormones must play a major role and a considerable amount of research is being carried out on this aspect. There are, however, numerous other factors that should be taken into account. We can note, for example, that locusts seem to be attracted by low atmospheric pressure which presages rain and leads to a temporary appearance of vegetation over desert regions. The phase theory, although considerably advancing our understanding of locust migrations, has by no means supplied all the answers.

We can now turn to the question of butterflies. We have already seen that, in spite of their apparent fragility and the short life-span usually attributed to them, they should in fact be included in the ranks of the great migrants. The methods available for studying their migrations have not been as precise as those used for other animals and it has not always been possible to trace migratory routes with the same precision. The bulk of the work has been accomplished by direct observations. A species appears in a particular place, having disappeared from another place, and one concludes that it has taken a particular direction and accomplished the journey at a particular speed. To verify these hypotheses it has been necessary to adopt marking techniques in keeping with the anatomy of such small creatures. In the earliest experiments the information was quite simply written in red ink on one of the wings. Later on, certain English entomologists stuck on to the wing of these unfortunate guinea-pigs a small piece of paper bearing a number and the legend 'London Zoo'. In recent times more refined techniques have been developed, the wings being dipped into dye, or clipped in some way, or a series of coloured dots painted on the wings – this technique has been used successfully with tsetse flies and the use of luminous paint enables their movements to be followed at night. But even the rather barbaric earlier marking methods produced interesting results which have proved the existence of migratory movements covering 100 miles or more.

The Painted Lady (*Vanessa cardui*) should probably be considered the queen of butterfly migrants. It is a species with pinkish-orange wings set off with patches of black and white and, apart from South America, it is found almost everywhere, in Europe, Australia and Asia, from the Arctic Circle to southern Africa.

The eggs are laid in the deserts of North Africa. They hatch out after the winter rains have produced a fresh but ephemeral burst of greenery on which the caterpillars feed. After a number of moults the caterpillars transform into chrysalids and shortly afterwards the butterflies emerge. The following is an account of the emergence of the butterflies by Sydney Skertchley in March 1869 near Suakin on the Red Sea:

Our caravan had started for the coast, leaving the mountains shrouded in heavy cloud, soon after daybreak. At the foot of the high country is a stretch of wiry grass, beyond which lies the rainless desert as far as the sea. From my camel I noticed that the whole mass of the grass seemed violently agitated, although there was no wind. On dismounting I found that the motion was caused by contortions of pupae of *V. cardui*, which were so numerous that almost every blade of grass seemed to bear one. The effect of these wrigglings was most peculiar, as if each grass stem was shaken separately – as indeed was the case – instead of bending regularly before the breeze. I called the attention of the late J. K. Lord to the phenomenon, and we awaited the result. Presently the pupae began to burst, and the red fluid that escaped sprinkled the ground like a rain of blood. Myriads of butterflies limp and helpless crawled about. Presently the sun shone forth, and the insects began to dry their wings: and about half-an-hour after the birth of the first, the whole swarm rose as a dense cloud and flew away eastwards towards the sea. I do not know how long the swarm was, but it was certainly more than a mile, and its breadth exceeded a quarter of a mile.

This is the way that the spring migration starts. The Mediterranean is reached by March or April, and the north of France by May. During the journey there is frequently a second breeding and the young join with their parents, or replace them, in the migration. In some years the Painted Lady crosses the Channel and may even leave England for more northerly latitudes.

The return migration, which has been observed in autumn over the Mediterranean and the west coasts of Africa, brings the Painted Lady back to tropical Africa once more.

Other migratory flights have been regularly seen over the Red Sea, around the Ukraine and over Pakistan at considerable heights (17,000 feet). These migrants are also known in North America. They pass the winter in the semi-desert regions of Mexico, and, as in Europe and Asia, they move northwards in the spring in gigantic swarms of hundreds of millions of individuals, reaching as far north as Canada. A return migration must exist but no clear evidence for it has yet been found. It is, in fact, one of the peculiarities of butterfly migrations that there is a great disparity between the two migratory phases. In most cases one does not see the return migration which would complete the life cycle. One wonders, therefore, whether this is not a form of invasion without return and not a true migration at all.

There are, however, certain instances that suggest an answer. In the Monarch butterfly, which we shall discuss shortly, the insects fly north either singly or in pairs. Now the Monarch is a very distinctive and well-known butterfly in the United States, so it is quite possible, even probable, that in this and other species a return flight exists but is not noticed partly because the return is made so much more discreetly and partly because the numbers have by that time been depleted.

The life cycle of the Monarch (*Danaus plexippus*) is closely tied to a particular plant, the Milkweed (*Asclepius* species), on which it lays its eggs and which forms the principal food of the caterpillars. In a month the caterpillars have reached their maximum size and after a week as a chrysalis the large and very beautiful butterfly emerges. It has a wingspan of about four inches and the orange-red wings are veined in very dark violet and edged with a speckling of white dots. There are two independent populations, one in the north and the other in the south of the American continent. From America, the Monarch managed to reach the Hawaiian islands in about 1850 and was able to breed there because of the presence of the milkweed plant. The Monarch is thus on a par with that other famous migrant, the Pacific golden plover (see page 90) which also travels no less than two thousand miles to reach Hawaii. Pressing on westwards the monarch has even colonized Borneo and New Zealand. Its movements to the east have been less profitable because, having reached Europe, it has not been able to establish itself because of the absence of the milkweed.

One wonders whether it has been able to cross the oceans under its own power or whether this has happened accidentally through man's agency. The rather recent colonization of Hawaii would seem to suggest the latter, but Monarchs have been observed some 600 miles from the English coast. This shows that some of the butterflies that have been blown out to sea as they migrated along the eastern coast of America are able to survive until they reach the British Isles. The problems posed by the introduction of species into countries where they formerly did not occur can be very serious. Quite often the introduced species finds itself free of its natural enemies and begins to proliferate without check, thereby causing a complete upset in the natural balance of species. One striking example of this was the unintentional importation into France of *Phylloxera*, an aphid which had achieved some sort of balance with American vines but for which the European vines had no resistance. It spread rapidly and the entire French wine and grape industry were faced with ruin until an American species of vine was introduced resistant to the aphid. The American grey squirrel is a good example of a species introduced into England with harmful results.

Apart from these airborne exploits, the Monarch carries out true annual migrations in America. The populations of North and South America are quite separate and do not mix, each having its own migration pattern. In North America the species of milkweed on which the larval Monarchs feed flourish in a relatively cold climate and thus cover a vast area, from Canada to the north of Mexico (Fig. 44). In September, when the first frosts are felt, the dispersal begins. The ranks of the migrants gradually swell until eventually they form a veritable flood of butterflies that travel night and day, sometimes very high in the sky, across the United States from north to south. When they arrive at some suitable place they settle in vast numbers on the branches of evergreen trees and hibernate for several months. The hordes of butterflies on the trees are sometimes so dense that the branches actually break with the weight of them. Year after year they occupy particular sites and these are well known and are shown off to tourists. In spring they make their way quickly northwards, singly or in small groups but not forming the vast swarms of the autumn migration. Copulation takes place at this time and at the end of the journey the ripe females lay their eggs on the milkweed, thus completing their life cycle.

In South America to the south of the Amazon there is a subspecies of the Monarch (*Danaus plexippus erippus*) which resembles that of North America but lacks the black bar at the lower edge of the fore wings. It shows a classic pattern of migration, moving away from the Equator in spring and towards the Equator in autumn.

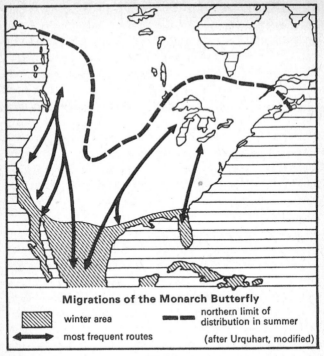

Migrations of the Monarch Butterfly

winter area — northern limit of distribution in summer

most frequent routes (after Urquhart, modified)

Fig. 44

The migratory habits of two butterflies have been given in some detail here, the Monarch and the Painted Lady. There are, however, many other migratory butterflies and a few can be mentioned here.

The Large Cabbage White (*Pieris brassicae*), as well as other members of the 'white' family (*Pieridae*), also have quite a complex migratory life cycle. The large cabbage white, perhaps the best known of all butterflies in England, is equally abundant in western Europe, particularly when immigrants add to the native populations. In England this is often the case in July and August, clouds of them coming from Scandinavia and fluttering down in great numbers after their North Sea crossing. In India the same species undertakes huge movements between the Himalayas and the plain of the Ganges.

Another, but much smaller, butterfly that deserves admiration for the size of its migrations is the Long-tailed Blue (*Lampides boeticus*), a small blue butterfly with characteristic hair-like projections from its hind wings. Almost world wide in its distribution it makes long journeys and crosses seas and mountains to reach its destination. It has been seen at heights of over 7,000 feet over the Pyrenees, and in the Himalayas is believed to make regular migrations up the slopes during the hot weather. It is not the only butterfly to attain great heights, however. The cabbage white crosses the Alps and numerous other species ascend the Pyrenees and then flutter down to the warmer climate of Spain. In Africa, some species speed across the continent, from Egypt to the Cape of Good Hope, and in South America gigantic flights congregate, occasionally made up of several different species. In fact, migratory butterflies are found everywhere.

Some nocturnal moths are also migratory, and the Death's-head Hawkmoth (*Acherontia atropos*) is an example of one that reaches as far as England and the Scandinavian countries, having originated in North Africa and southern Europe. It is even recorded from Iceland.

Green, blue, red or brown – dragonflies with their superb flying abilities are well adapted to migratory flights. Usually, one sees them hovering and for a moment one can glimpse their colours before they are off in a flash to the other side of the stream or pond. Sometimes, however, they give up their solitary habit and congregate in groups and then depart for some more or less precise destination. Europe has witnessed vast invasions, especially of the Four-spotted Libellula (*Libellula quadrimaculata*) and the Broad-bodied Libellula (*L. depressa*). The former have been recorded in millions in Heligoland, appearing shortly before thunderstorms and later disappearing. The journeys of dragonflies may be measured in hundreds of miles, as for example the invasion of *Sympetrum striolatum* into Ireland in 1947, the individuals almost certainly having come from Spain and Portugal after a flight over the sea of more than 500 miles. The flights of another species, *Panatala flavescens*, are even more incredible, for this dragonfly was seen in 1896 about 300 miles from the Cocos-Keeling Islands and 900 miles from the nearest coast of Australia. In America, too, there are often vast swarms of dragonflies, sometimes flying in the company of butterflies.

The Seven-spot Ladybird (*Coccinella septempunctata*) – known as *La Bête au Bon Dieu* in France – is also a hardy traveller in spite of its rather delicate appearance. It regularly oscillates between its summer and winter quarters. In 1952 there was a huge migration along the coast of south Lincolnshire, the high-tide mark being reported pink with millions of these insects along forty miles of the coast. In June 1925 there was a great movement of the Two-spot Ladybird (*Coccinella bipunctata*) over the northern half of England. The Eleven-spot Ladybird (*Coccinella undecimpunctata*) forms truly prodigious swarms under certain conditions; in 1939 Professor Oliver saw the remains of a part of such a vast flight along the shores of northern Egypt and he estimated that there were 4,500,000,000 insects, or about 70,000 per foot of shore.

Then there are the flies. One cannot but mention the Housefly (*Musca domestica*), that most domestic of animals. Who has not wondered where houseflies go in winter? After so many and such varied examples of migrants, it will come as no surprise to hear that other species of flies have been observed travelling over the sea, in the company of butterflies or ladybirds, and some like the Hoverflies (*Syrphidae*) have been observed crossing passes in the Pyrenees at 7,000 feet and the Himalayas at 12,000 feet.

Even greenfly are not immune to the migratory habit, the winged females taking advantage of the warm summer breezes to disperse from their breeding areas. Sometimes they are so numerous that clouds of them darken the sky. During the summer, these insects reproduce parthenogenetically, that is to say the females produce young from eggs that have not been fertilized by the male. This virgin birth continues for successive generations, but each time the progeny, although sometimes winged and sometimes wingless, are always females. Only in October do the males at last appear and they fly with the winged females to some kind of woody plant or tree where the females produce wingless daughters by virgin birth and these finally mate with the males to produce fertilized eggs.

Greenfly are not the only insects to take advantage of winds in order to disperse. The aerial plankton – for that is the name given to the fantastic numbers of small insects that float around in the air – travels passively at the whim of the winds and can cover distances of up to about sixty miles a day. This is not, however, the rule in migratory insects, for most species do not in fact rely on the winds and it is not uncommon to see butterflies struggling against the wind (locusts, as we have mentioned earlier, are in some cases controlled by the winds that meet at the Intertropical Convergence Zone).

One can thus speak of active and passive migrants, the former migrating under their own power, with the direction of flight determined by themselves. This leads one to ask what means insects have of orientating during their flights. A great deal of research is going on at present on this question. While many solutions have been proposed, it still remains for theories similar to those proposed to account for orientation in birds to be proved in the case of insects. There are, however, a number of conclusions that seem incontestable. First of all, many insects are known to use the sun as a means of navigation and this was very clearly shown by experiments carried out by one of the most ingenious of experimenters, Professor Karl von Frisch.

Von Frisch trained hive bees to come to their food (sugared water scented with lavender) which was set on a table some 200 yards to the west of the hive. When the bees were trained, the hive was transported several miles away to a place that was quite different in appearance from the first and in such a manner that the bees were unable to use their visual memory in the tests that were to follow. Then the same feeding apparatus was set up but this time on four tables spaced 200 yards to the north, south, west and east of the hive. The result was that, even though the territory traversed by the bees was different, and even though the entrance to the hive might be set facing south instead of east, the great majority of bees came to the table to the west of the hive and, moreover, this occurred whatever time of day was chosen.

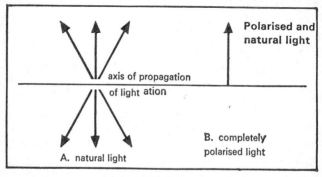

Fig. 45

This experiment conclusively showed that bees, like many other species of animal, were capable of orientating by means of the sun. Von Frisch believed that it was by analysing the pattern of polarization of the sky that the bees were able to achieve their goal. In certain respects, light can be considered as a wave that travels through the air at a speed of 186,000 miles a second. This wave can be defined by a vector, that is to say by reference to its direction, amplitude and the plane in which it vibrates. Theoretically, natural light is symmetrical and its planes of vibration are in all directions but in practice light from a blue sky is partially polarized and the pattern of polarization depends on the position of the sun. Thus, seen through a polarizing screen, certain parts of the sky look darker than others (when the lenses of two pairs of polarizing sunglasses are placed together and slowly rotated, the light is entirely cut out when the planes of polarization are exactly at right-angles to each other; a single pair is merely eliminating light vibrating in one plane). Thanks to the very

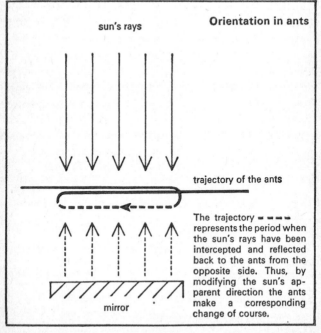

Fig. 46

complex structure of their eyes, the bees are able to analyse polarized light in the sky and thus to navigate.

Ants also use the sun to navigate. A very simple experiment, described and illustrated in Fig. 46, shows that this is so.

Other insects appear to be sensitive to both electro-magnetic and to gravitational forces (theoretical physics have shown that, sensitive to these two forces, insects must also be sensitive to the inertial forces of spin). Thus a cockchafer has been shown to orientate relative to the distribution of masses around it when it was placed in a graduated circular pen. One very spectacular experiment was carried out. A cockchafer was placed in such a graduated circular pen and two opaque containers were placed on either side of the circle. In the first instance these were empty and the animal orientated in a particular direction. On the second occasion the containers were filled with a weight of 44 lb. each. The insect then modified its orientation. This was an absolutely astounding result, for one must envisage the minute gravitational effect that a mass of 44 lb. must exert on an insect weighing only one gramme and at a distance of about four feet. The effect between two such masses at this distance is of the order of 0·00000001 gramme force. It remains, however, to determine the role that such astonishing abilities play under real migratory conditions. In fact this is one of a number of problems that await solution.

Amphibians, Reptiles, Crustaceans, etc.

Most toads and frogs undertake a very characteristic migration. Each year populations that are spread over a relatively wide area congregate in special places to breed. Such breeding places are very precisely delimited, and if one dams up a particular lake or drives a road through the area, the toads are annihilated since they refrain from breeding in what appear to be quite similar lakes near by. In Africa one finds that at very precise times, just before the rainy season, innumerable cohorts of frogs make their way to the lakes where they breed, guided by an extremely well-developed sense of direction. In fact, six species of frogs are now known to orientate using celestial cues (moon, stars, sun) in order to move towards or away from a familiar segment of shore-line. In one species, *Rana catesbeiana*, Denzel Ferguson showed that the frogs orientated both by day and by night but not under cloudy conditions and that they orientate towards natural shore-lines in the spring and autumn but not in the summer during the breeding season (when they would normally be at the shore).

Reptiles, which are often shy and solitary during most of the year, become gregarious during the breeding season in some species. Snakes, for example, often have a very precise place in their general hunting territory where they establish a traditional breeding site.

Among the polychaete worms there is the palolo worm (several species), which shows a form of behaviour that just falls within our definition of migration. This worm is found in the warmer waters of the Pacific and it has a very definite annual cycle which in this case is regulated by the moon. It is, in fact, in the moon's last quarter during October and November that the worms breed. At this time they split in two and the hind segments (orepitokes) containing eggs (or in males, containing the spermatozoa) rise to the surface, sometimes in such vast numbers as to colour the sea. The influence of the moon on marine organisms is fairly widespread, either direct or through the intermediacy of the tides. Thus, there is a fish, the Grunion (*Leuresthes tenuis*), whose life cycle is very closely linked with the moon. For a few days that follow the full moons of March to June (i.e. at the time of the high spring tides) these fishes swim up the shore as far as possible to bury their eggs in the sand. The eggs remain dry but hatch out within three minutes when the next high spring tide reaches them, and the larvae then escape to the sea.

The Sandhopper (*Talitrus saltator*) also has some intriguing gifts. These little crustaceans, which occur in millions on sandy beaches in Europe, are able to wander some distance from the shore but can return without hesitation and by a direct route. If one takes them inland in a jar, one finds them turning towards the sun and then orientating very exactly in the direction of the shore. The precision of the response becomes less during the course of a few days if one does not recharge the animal, so to speak, by letting it familiarize itself with its own home ground. Two Italian workers, L. Pardi and F. Papi, studied these animals along the Adriatic coasts of Italy and reported the following results.

If one takes these animals far from the coast, they will in this case go eastwards and keep this direction until they reach the Italian coast. But if they are taken to the other side of the country, to the Mediterranean shore, and then released, they still head eastwards and thus go farther and farther from the shore. They are able, therefore, to keep to a constant direction but they cannot orientate to a particular point. By experiments similar to those carried out on other animals, Pardi and Papi were able to show that these animals orientate not only by means of the sun compass but, to the surprise of these two workers, also by means of the moon. This, if true, is very surprising because the movements of the moon are very much more complex than the apparent movements of the sun and the amount of compensation needed in order to hold to a straight course is much greater. As yet, rather few animals have been found to use the moon as a reference point.

Marine crabs, in spite of their rather slow method of walking and swimming, are capable of surprisingly long migrations, and distances of over 100 miles have been recorded. Generally, the females leave the shallow waters after breeding, leaving the more sedentary males behind.

Causes of Migration and Conclusions

We have given a sufficiently broad spectrum of animal migrations in various groups to be able to consider migrations as a whole. The most important consideration is the fact that life itself is synonymous with rhythm, for every aspect of nature is permeated with some kind of oscillation, some kind of pulse. It is not surprising, therefore, that one should find rhythms in the behaviour of animals. From the carp that oscillate from one part of the lake to the other for breeding, to the albatrosses that circle the globe, there is probably no animal which does not show some kind of rhythmic behaviour.

The most important aspect of migratory behaviour is that it is cyclic: the movements are neither haphazard nor fortuitous. Although there are many examples of spectacular migrations, judged that is by the distances covered, it is wrong to think that only animals that make long journeys are necessarily migrants. Sedentary animals, that is to say animals that do not make long journeys but nevertheless move around their area of distribution, are in fact in the majority. Thus, there are over 1,000,000 different species of insects but only a small part of these move any great distance. It is only amongst the birds that we find a majority of migrants.

The question of the causes of migration has given rise to a plethora of theories and hypotheses that stem from inadequate experimental data. One theory may fit one set of experimental facts but is not able to explain another set of facts, so that as yet we lack any firm principles. However, at one level of discussion one can say that migration as we see it in the animal world is an essential factor in the maintenance of biological equilibrium. Not, it should be stressed, that this is a *cause* of migration, but that equilibrium results from migration (although another form of equilibrium would be achieved by different means were migration not to occur). Migration does, in fact, serve to limit the size of a population in a particular area (one out of every two swallows is lost during migration) and it must therefore contribute to the harmonious balance between the various elements in any natural environment. This is most apparent in the case of the lemmings and the springboks, that is in the case of migrations without return (i.e. invasions) in which the excess of the population is removed at a single stroke. This is seen very clearly in the Garden Warbler (*Sylvia borin*) which, having crossed the Sahara, does not stop in the rich parts of Angola but presses on further southwards to the deserts of Namib where life is much more harsh.

Not all cases are of this kind, for there are numerous migrants that actively search out the best conditions possible. Birds, for example, may follow temperature belts in which insects are hatching out, while mammals move with rainfall belts in their quest for better pastures. Equilibrium is, nonetheless, maintained. If garden warblers remained in Angola they would soon multiply and become too abundant for the resources, whereas if nightjars did not follow the areas of greatest insect density they would soon die out. One must conclude, therefore, that migration is not the *only* solution to the problem of the balance between animals and food resources, but that it is the one that exists at the present time.

With regard to birds, one can ask how the intercontinental migrations arose and the following hypothesis can be put forward. If a particular biological equilibrium is upset, perhaps as a result of climatic changes, this may lead a part of the population to colonize new areas. If, unlike the case of invasions, such colonization succeeds, then the birds may return periodically to their original habitat if conditions are not favourable throughout the year. However, the reverse may also happen. A cooling of the climate in the original area may drive the birds to a warmer country in the winter, the birds returning directly the warm weather returns. This would give rise to migrations to and from the breeding area.

Regarding the faculties of orientation that have been demonstrated in certain animals, we have seen that it would be unwise to attribute to such animals the ability to reason deductively. In reality, their perception seems to be spontaneous, that is to say immediate and intuitive in the way that we ourselves can in certain instances immediately grasp an impression either visually, audibly or as a pattern of thought.

Let us end on a rather mysterious note.

Some cats were taken from a town and placed in darkened containers and then transported some miles away after a most complicated journey full of detours and retracing of steps. There was no possibility that the cats could use their memory of the route for this experiment. The cats were then taken from the containers and placed in the centre of a large maze with twenty-eight exits.

The great majority of cats chose the exit that was in the direction of their home.

Bibliography

BISON
Garretson, M. S., *The American Bison*, New York Zoological Society, 1938.
Roe, F. G., *The North American Buffalo*, University of Toronto Press, 1952.

ELK
Murie, O. S., *The Elk of North America*, Wildlife Management Institute Publications, 1951.

CARIBOU
Boone and Crocket Club, *North American Big Game*, Scribner, New York, 1939.

SPRINGBOK
Schreiner, S. C. Conwright, *The Migratory Springbok of Africa*, 1925.

BATS
Allen, G. Morill, *Bats*, Harvard University Press, 1939.
Brosset, André, *La Biologie des Chiroptères*, Masson, 1966.

SEALS
King, J., *Seals of the World*, British Museum Natural History Booklet, 1964.
Maxwell, Gavin, *Seals of the World*, Constable, 1966.

WHALES
Norman, J. R., and Fraser, F. C., *Giant Fishes, Whales and Dolphins*, Putnams, 1948.
Slijper, E. S., *Whales*, Hutchinson, 1962.

BIRDS

Dorst, J., *The Migrations of Birds*, Heinemann, 1962.

Griffin, A., *Bird Migration*, Heinemann, 1965.

Matthews, G. V. T., *Bird Navigation*, Collins, New Naturalist Series, 1948.

Peterson, R., Mountford, G., and Hollom, P. A. D., *A Field Guide to the Birds of Britain and Europe*, Collins, 1966.

Thomson, A. L., *A New Dictionary of Birds*, Nelson, 1964.

FISHES

Hasler, A. D., *Homing of Salmon: Underwater Guideposts*, University of Wisconsin Press, 1966.

Netboy, A., *The Atlantic Salmon*, Faber & Faber, 1968.

INSECTS

Urquhart, F. A., *The Monarch Butterfly*, Oxford University Press, 1960.

Uvarov, B. P., *Locusts and Grasshoppers*, London, 1928.

Williams, C. B., *The Migration of Butterflies*, Edinburgh, 1930.

Williams, C. B., *Insect Migrations*, Collins, New Naturalist Series, 1958.

GENERAL

Carson, Rachel, *The Silent Spring*, Hamish Hamilton, 1963.

Corbet, G. B., *The Terrestrial Mammals of Western Europe*, Foulis, 1966.

Elton, C., *Animal Ecology and Evolution,* Oxford, 1930.

Fraenkel, C. S., and Gunn, D. L., *Orientation of Animals,* Oxford, 1930.

Graaf, J., *Animal Life of Europe*, Warne, 1968.

Heape, W., *Emigration, Migration and Nomadism*, Cambridge, 1931.

Lockley, R. M., *Animal Migration*, Arthur Barker Ltd, 1967.

Palmer, R. S., *The Mammal Guide: Mammals of North America north of Mexico*, Doubleday, New York, 1954.

Index

196

On The Track of Unknown Animals
Bernard Heuvelmans

At the beginning of the nineteenth century George Cuvier, the 'Father of Palaeontology', categorically stated that there was little hope that any large animals were still unknown. Since then have been discovered the largest species of bear and gorilla, a white rhinoceros, the pygmy elephant, the okapi, the fabulous Komodo dragon and dozens of other unusual animals. Bernard Heuvelmans, 'The Sherlock Holmes of Zoology', believes, in the face of official zoology, that there are more animals yet to be found. He examines the evidence for the spotted lion of Kenya, the Queensland marsupial tiger, the abominable snowman and other equally 'fantastic' animals.

'Dr Heuvelmans' original research beats any alleged thriller for enthralling excitement; incredible too, this is first-rate authoritative science written without jargon.'

DAILY MAIL

'This is the well reasoned thesis of a scientist who is a good deal more open-minded than many of his fellows ... Even if none of these questionable beasts should be discovered, which is unlikely, the author has documented their legends in a masterly and useful way.'

TIMES LITERARY SUPPLEMENT

Paladin